NATO ASI Series

Advanced Science Institutes Series

A series presenting the results of activities sponsored by the NATO Science Committee, which aims at the dissemination of advanced scientific and technological knowledge, with a view to strengthening links between scientific communities.

The Series is published by an international board of publishers in conjunction with the NATO Scientific Affairs Division

A Life Sciences B Physics	Plenum Publishing Corporation London and New York
C Mathematical and Physical Sciences D Behavioural and Social Sciences E Applied Sciences	Kluwer Academic Publishers Dordrecht, Boston and London
F Computer and Systems Sciences G Ecological Sciences H Cell Biology I Global Environmental Change	Springer-Verlag Berlin Heidelberg New York London Paris Tokyo Hong Kong Barcelona Budapest

NATO-PCO DATABASE

The electronic index to the NATO ASI Series provides full bibliographical references (with keywords and/or abstracts) to more than 30000 contributions from international scientists published in all sections of the NATO ASI Series. Access to the NATO-PCO DATABASE compiled by the NATO Publication Coordination Office is possible in two ways:

- via online FILE 128 (NATO-PCO DATABASE) hosted by ESRIN, Via Galileo Galilei, I-00044 Frascati, Italy.
- via CD-ROM "NATO-PCO DATABASE" with user-friendly retrieval software in English, French and German (© WTV GmbH and DATAWARE Technologies Inc. 1989).

The CD-ROM can be ordered through any member of the Board of Publishers or through NATO-PCO, Overijse, Belgium.

Series F: Computer and Systems Sciences Vol. 86

The ASI Series Books Published as a Result of Activities of the Special Programme on ADVANCED EDUCATIONAL TECHNOLOGY

This book contains the proceedings of a NATO Advanced Research Workshop held within the activities of the NATO Special Programme on Advanced Educational Technology, running from 1988 to 1993 under the auspices of the NATO Science Committee.

The books published so far as a result of the activities of the Special Programme are:

Vol. F 67: Designing Hypermedia for Learning. Edited by D. H. Jonassen and H. Mandl. 1990.

Vol. F 76: Multimedia Interface Design in Education. Edited by A. D. N. Edwards and S. Holland. 1992.

Vol. F 78: Integrating Advanced Technology into Technology Education. Edited by M. Hacker, A. Gordon, and M. de Vries. 1991.

Vol. F 80: Intelligent Tutoring Systems for Foreign Language Learning. The Bridge to International Communication. Edited by M. L Swartz and M. Yazdani. 1992.

Vol. F 81: Cognitive Tools for Learning. Edited by P.A.M. Kommers, D.H. Jonassen, and J.T. Mayes. 1992.

Vol. F 84: Computer-Based Learning Environments and Problem Solving. Edited by E. De Corte, M. C. Linn, H. Mandl, and L. Verschaffel. 1992.

Vol. F 85: Adaptive Learning Environments. Foundations and Frontiers. Edited by M. Jones and P. H. Winne. 1992.

Vol. F 86: Intelligent Learning Environments and Knowledge Acquisition in Physics. Edited by A. Tiberghien and H. Mandl. 1992.

Intelligent Learning Environments and Knowledge Acquisition in Physics

Edited by

Andrée Tiberghien

C.N.R.S. - IRPEACS
93 Chemin des Mouilles, B.P. 167
F-69131 Ecully Cedex, France

Heinz Mandl

Universität München
Institut für Empirische Pädagogik
und Pädagogische Psychologie
Leopoldstrasse 13
W-8000 München 40, FRG

Springer-Verlag
Berlin Heidelberg New York London Paris Tokyo
Hong Kong Barcelona Budapest
Published in cooperation with NATO Scientific Affairs Division

Proceedings of the NATO Advanced Research Workshop on Knowledge Acquisition in the Domain of Physics and Intelligent Learning Environments, held in Lyon, France, July 8–12, 1990.

CR Subject Classification (1991): K.3.1, I.2.6, I.2.1, J.2, J.4

Additional material to this book can be downloaded from http://extra.springer.com.

ISBN-13: 978-3-642-84786-8 e-ISBN-13: 978-3-642-84784-4
DOI: 10.1007/978-3-642-84784-4

Softcover reprint of the hardcover 1st edition 1992

Typesetting: Camera ready by authors
45/3140 - 5 4 3 2 1 0 - Printed on acid-free paper

Preface

The NATO workshop "Knowledge acquisition in the domain of physics and intelligent learning environments" was held in Lyon, France, July 8–12, 1990. A total of 31 researchers from Europe (France, Germany, Greece, Italy, Portugal, and the U.K.), the U.S.A., and Japan worked together.

This proceedings volume contains most of the contributions to the workshop. The papers show clearly the main directions of research in intelligent learning environments. They display a variety of points of view depending on the researcher's own background even when a single domain of teaching, namely physics, is considered.

We acknowledge the assistance of Michael Baker, who was responsible for reviewing the English of the contributions.

February 1992

Andrée Tiberghien
Heinz Mandl

Table of Contents

Introduction

The workshop "Knowledge acquisition in the domain of physics and intelligent learning environments" brought together researchers from different backgrounds in physics education (didactics of physics), cognitive psychology, and artificial intelligence, to study intelligent learning environments in the domain of physics in terms of four different aspects.

1. Teaching situations and physics knowledge

Intelligent learning environments concerned with experimental sciences use real or simulated experiments. The choice of the concrete situations incorporated in these experiments appears to be very crucial because:

– They should be coherently related to the concepts to be taught. The question of the importance of an epistemological analysis of teaching situations was raised.

– They depend on the choice of learning theory, where emphasis is put on different aspects of knowledge.

– The modelling of these concrete situations should be compatible with the other models involved in the knowledge to be taught and in learner models. For example, the question of the methods of qualitative physics was raised.

Intelligent learning environments in this domain raise also the question of knowledge representation, which is crucial if we want to model problem solving.

2. Different approaches to student modelling

Two orientations were mainly discussed, differing in their implied learning theories and epistemological position:

– The syntactic aspects of the knowledge involved, that is, the relations between the physical variables used to describe the experiment, are emphasized. In this case, the approach to student modelling used is not intended to be specific to the physics knowledge involved.

– The semantic aspects of knowledge, that is, the relationships between objects and events involved in the experiments and physical variables, are emphasized. In this case, the student model comes from an analysis of conceptions held by learners of the tasks.

These two approaches are often characterized by the nature of the simulated situations. In the first case, they are made in terms of the symbolic representations specific to a model in physics. In the second case, they are made in terms of diagrams representing the objects in the environment.

These orientations form the basis for extensive collaborative work.

NATO ASI Series, Vol. F 86
Intelligent Learning Environments
and Knowledge Acquisition in Physics
Edited by A. Tiberghien and H. Mandl

3. The interaction learner/learning environment

There appeared to be a large measure of agreement on the importance of guidance in the interaction between the learner and the learning environment. Two aspects emerged:

– The type of relevant information to be collected and analyzed from the learner's actions for modelling;

– The type of decision to take in the intervention in the environment.

For example, in a negotiated dialogue the question arises of which reasons should be used to decide whether the learner or the environment makes the decision.

4. Design of learning environments

This was a major aspect for some participants even though it cuts across the previous ones. Taking into account the different backgrounds of the participants, the research involved in the design of learning environments ranges from questions about students' conceptions and difficulties in learning specific aspects of physics to questions about modelling students from their actions at the interface.

We have organized this book into four parts to reflect the four main aspects of the workshop.

The first part deals with teaching situations and physics knowledge. Balzano et al. outline possible roles of a learning environment in teaching physics. Coutois discusses knowledge representation from an artificial intelligence perspective. Hron discusses expert problem solving from didactic and psychological points of view. Tisseau presents an artificial intelligence system for modelling problem solving.

The second part deals with student modelling. Bliss and Ogborn propose steps toward formalizing a commonsense theory of motion, Grob et al. report an expert system test of how students structure knowledge, and Vosniadou argues that learning involves reorganization of conceptual structures. Mandl and Hron discuss cognitive approaches to student modelling, Hron et al. report the development of an automatic diagnosis system on knowledge acquisition and on misconceptions, and Reimann describes an interactive simulation environment for learning by hypothesis testing.

The third part presents papers dealing with interaction. Baker presents a method for analyzing students' collaborative explanations of phenomena, and Tiberghien analyzes interfaces from the points of view of epistemology and didactics.

The fourth part presents different approaches to designing a learning environment in the domain of physics. Borghi deals with reconstruction of historical experiments, Caillot presents two case studies from mechanics, and Loureiro describes a constructivist strategy for teaching electricity. McDermott discusses how research on the teaching and learning of physics can guide the design of intelligent learning environments. Finally, Thornton presents evidence for improved learning by students using microcomputer-based laboratory materials compared to those taught in a traditional manner.

1. Teaching Situations and Physics Knowledge

Introductory University Courses and Open Environment Approaches: the Computer as a Multi-Role Mediator in Teaching/Learning Physics

Emilio Balzano[1], Paolo Guidoni[2], Maria Moretti[3], Elena Sassi[1], and Gabriella Sgueglia[4]

[1]Dipartimento di Scienze Fisiche, Universita' di Napoli, Mostra d'Oltremare pad.20, I-80125 Napoli, Italy; E-mail DAEES @ VAXNA1.INFN.IT.BITNET

[2]Seminario Didattico della Facolta' di Scienze, Universita' di Napoli, via Rodino' 22, I-80138 Napoli, Italy

[3]Istituto Tecnologie Didattiche, CNR, via all'Opera Pia 11, I-16145 Genova, Italy

[4]Biblioteca Software Educativo, CISED, Mostra d'Oltremare pad.17, I-80128 Napoli, Italy

Abstract: Teaching/learning of university introductory physics is a specific context for testing computers' roles in interfacing individual thinking to disciplines' formalized structures. An experiment is in progress at Naples University. We outline features and problems of an open environment approach to computer-supported, laboratory-based activities. Three *modes* of teaching/learning articulate the activities: *qualitative* approach to formalization; *quantitative* study of *simple* situations; *quantitative* approximate approach to *complex* situations. We suggest that well planned computer use and substantial requalification of the teaching effort can extensively and qualitatively improve the average level of understanding.

Keywords: multi-role computer, open approach, university physics, on-line transduction, correlated representations, learning strategies, theory-practice

1. Introduction

It is now commonly accepted that knowing and learning are extremely complex processes. The methods to represent them have rapidly envolved in recent years, also thanks to use of computers and to computational models of thinking. Cognitive theories, however, as sophisticated as they can be, still appear unable to adequately describe and address what happens in complex teaching-learning interactions.

Knowledge is based on thousands of elements. Many of them are not easy to define and complexity appears in the *links* between specific (disciplinary) topics and general (qualitative) knowledge, which contribute as basic ingredients to the formation of students'

NATO ASI Series, Vol. F 86
Intelligent Learning Environments
and Knowledge Acquisition in Physics
Edited by A. Tiberghien and H. Mandl

understanding of science. In fact it is now commonly verified that students tend to interpret what they are exposed to in the frame of their functional knowledge systems. Teaching and learning physics, in particular, has to do with a merging- and- fitting process involving different structures: from personal cognitive strategy and style, experience in the context, commonsense, etc ..., to interwoven structures of physics and mathemathics as disciplines, to structures of communication in explaining-understanding interactions .

On the research front, for example, the *alternative conceptions* have been (and are being) studied in physics education research in a rather extensive way and on several topics, mechanics and heat in first place [1]. Such conceptions appear to be widespread, reasonably stable and quite robust; students continue to use them in spite of long exposures to physics courses, while teachers (and text books) are not immune from them. On the teaching side, in the meantime, it begins to be accepted that, at all levels and across all teaching schemes, the *interference* of what is taught with the student's functional knowledge system is indeed a crucial element. However there is still a diffuse resistance to quit the model that simply providing students good quality *accessibility* to subject matter, in the format of best logical organization and complete exposition, is enough to rouse learning.

In the following we briefly discuss some indications emerging from an educational research, still in progress, at the level of University Introductory Physics Courses (UIPC). We focus here on general issues, the description of the detailed results will be presented in a separate paper.

2. Theoretical Framework

Two introductory remarks are worth to be made about the very concept of teaching research.

- From a methodology point of view, a difficult (and partially ambigous) task is faced when planning and developing teaching research, i.e. educational research aimed at optimizing cultural transmission, in a given discipline. Clearly many results and models from cognitive psichology can contribute to such an endeavour: from natural conceptions to semantic networks' phenomenological * results, from piagetian to computational or connectionist theories, from motivation to peer-interaction studies. However, by definition, teaching research is compelled to be a real-time, real-place, real-conditions process and to preserve reasonable efficiency and efficacy of teaching while changing teaching modes, while getting feedbacks from induced (partial) changes.

---

* We currently use words like "phenomenology" and "phenomenological" with a variety of connotations, centered around the meaning of "observed or apparent object, fact or occurence". Please note the difference with the assumptions of "phenomenological" philosophycal currents

- From the content point of view teaching research, to be productive, has to deeply involve the overall conceptual structure of the subject matter (an area of Physics, in our case) at the proper formalization level. Relevant to cultural transmission, are, in fact, constructive abilities, to transfer and to correlate among different topics and contexts: at least as much as contextual, problem-triggered performances. This implies, at research level, an intrinsic non-separability among general (cognitive) and particular (disciplinary) aspects of involved competences and of possible results. Furthermore deep cognitive changes usually occur by long term, diffuse modes; that is why teaching research is so hard and fatiguing, often so intriguing.

Two main groups of hypotheses are subsumed into our theoretical framework, defining at the same time a general cognitive model, a guide to activity planning, a hyerarchy of teaching objectives, the criteria to interpret results and to elicit feedbacks. Such a research rationale actually represents the convergence between two different research lines: on educational use of computers in physics teaching and on investigating and modelling cognitive dynamics in physics learning.

For the sake of semplicity we'll therefore schematize our assumptions - objectives - criteria under two headings: Open Environment Approaches (OEA), from the point of view of computer's use; Integration (in context and abilities), from the point of view of learner's performance.

2.1. Open Environment Approaches

While describing research on computer use in teaching-learning we like to recall that questioning about "how much user's control ?" is a key issue in designing and using any educational hardware -software- courseware in physics education.

Different underlying cognitive models, no matter how much explicit or implicit, can lead in fact to approaches and materials defined by very different characteristics. Pedagogical interventions referring to constructivistic learning models tend for example to propose flexible types of computer use, offering the learner a friendly set of functionalities/ tools/ commands suited for working on a specific task, without any pre-embedded pegagogical strategy . At the other extreme, the interventions inspired by behaviouristic learning models tend to support directive computer uses guiding the learner along a predefined path toward precise goals. Actually there is a continuum range of possibilities, user's control decreasing when moving from OEA to directive tutorials [2-6].

Of course, the choice of the *best and most suitable* computer based activities depends on the specific educational situation as well as on the pedagogical objectives and the available resources. In particular, Introductory Physics Courses, even at University level, face resistant, widespread difficulties in the learning process, already documented by the research. [1, 7,8].

A *good* UIPC course should therefore help the student to get high level cognitive skills and abilities: expressing one's own mental frameworks or hypotheses and checking for their consequences; building a personal pedagogical path, self-evaluating and modifying it; proposing reasonable problems/questions and solving them, etc... appear in fact normally not enough mastered, yet necessary to meaningful understanding. OEA types of software seem to be very useful tools in helping to develop such capabilities: as the lack of pre-embedded strategies assures an intrinsic flexibility, the learner can control the pedagogical path and explore its alternatives, while computer's uses are easily integrable with no-computer activities at different levels of complexity, with looser integrability to the curriculum content. Actually, from our experience, to aim at high level cognitive skills (which are not confined to a specific discipline, but which are not easily measured) pays also on general learning levels, and can often activate unmotivated students or teachers.

Sometimes, however, OEA are criticized as *elitist* pedagogical strategies selecting "best" students and teachers. In fact the risk exists, when using OEA, of margining the less autonomous students and teachers, was OEA left in a pedagogical vacuum.

On the other side, one has to cope with a widespread increasing demand for guidance, both from students' and teachers' side. From our experience the suggestion emerges not to enforce strictly definite paths but to propose a multiplicity of examples, hints, stimulated choices, loosely guided sequences of activities, privileging the *network* model to the *ladder* one. In other words, to fully exploit the cognitive potentialities of the OEA an external, definite and effective framework has to be provided, both to teachers and students, to ensure that the learning process will eventually converge toward lasting conclusions. Such a framework, by integration of teacher's competence to supportive written material, can effectively elicit of branching points and critical nodes, support peers learning, address integration with the other activities, making possible *intelligent computer uses* even without direct AI applications .

Recently both learner centered pedagogical strategies and OEA computer uses are receiving increasing attention in physics education research. Such a shift in focus is favoured by several factors, such as the growing awareness of the roles that different mental schemata and learning strategies can play in the learning/teaching process; the improvements in the quality of laboratory activities afforded by on-line computer techniques; the need for meaningful scientific education to reach larger numbers of students; the increasing availability of materials inspired to OEA and belonging to different cathegories of software (knowledge based software, content independent general purpose tools, on-line transduction, multimedia environments, Intelligent Tutoring Systems); and so on. Within such a context, quite naturally it comes out that several problems arising in the educational research using

OEA are strictly related to key questions faced in designing and testing Intelligent Tutoring Systems (ITS). Among these: trying to identify and to induce peculiar features of "good" learning situations; enhancing the stimuli for both the differentiation and the convergence of personal learning processes; using the OEA itself to observe, to identify, to control and to address the reasoning strategies at work in individual students.

Altogether, computer supported OEA allows to expand drammatically the affordable situations, both from research and teaching-practice points of view. On the other hand it is worth noting that the proposal to diffuse OEA can be indeed an *explosive* one from the point of view of teaching organization, as it demands for more skilled and more autonomous teachers, finalized restructuring of curricula, irreversible processes of transformation of teachers' role in a redesigning of school activities and attitudes.

2.2. Integration

We will now schematically outline, under the heading of three clusters of objectives-assumptions-criteria, some features of our intervention, concerning physics [7-8], understanding [11-14]. and understanding of physics [7-10, 14-17). Across any cluster we stress the notion of integration: underlining the necessity of a coherent merge and reciprocal fit among different aspects of the usual teaching-learning situations, in order to achieve meaningful convergence for the explaining-understanding interactions [14].

2.2.1. Integration between the different contexts-attitudes-abilities-approaches-knowledges by which is normally articulated the practice and competence of a physicist. In particular, some aspects appear to be critically, but not transparently at all, linked to each other to condition understanding (i.e. constructive autonomy), such as:

i) The theoretical (conceptual-logical-mathematical) network according to which a physics' content area (i.e. mechanics) can be actually structured at a given level of understanding (i.e. UIPC); ii) The problem-solving activities, by which abilities to coherently shape-up and elaborate conceptually schematized situations are developed; iii) The laboratory activities, pursuing analogous but partially independent objectives in purposeful measurement contexts; iiii) The computer supported activities, where hardware and software can play critical and variable roles as real cognitive mediators (not only as flexible and quick performers) in the reciprocal fit of the previous aspects.

2.2.2. Integration between different computer roles within the processes of physical interpretation [15]. In particular we stress that totally new paths are allowed to the understanding-learning dynamics by interweaving of computer performances because:

a) On-line acquisition of data potentially subverts many aspects of traditional learning of Physics: this way, in fact, variables can be directly and operationally selected by the transducer, represented just in their varying, easily elaborated, represented and correlated; furthermore the course-shape acquires an independent relevance and status, as compared to the point-value, with implications ranging from perceptual insight to formal error analysis. A "variational" approach to the understanding of a physical situation (i.e. to the grasping of its internal structures) becomes for the first time possible and profitable, by intelligently modfying parameters and conditions of the experiment without the burden of time-wasting measurements.

b) Simulation and modelling by flexible and transparent programming can guide meaningful paths from acquaintance to mastery in the necessary relationships between physical concepts and mathematical representations, another crucial step in the appreciation of physics ability to interpret facts. At any level of mathematical competence, furthermore, <whatever can be coherently written down into an equation, that can be solved> by the computer, then represented and explored in its meanings; while the fit between actual and model data can be immediately visualized, and quantitatively evaluated.

c) Multirepresentation (change of scales, but also of variables, to explore and to show forms and correlations) allows for the gradual realization that physical meaning is stored just in forms and shapes, and in correlations among forms and shapes. Again the speeding-up allowed by the computer entails a deep, qualitative change in learning strategies, by making accessible a real multiplicity of alternative presentations of data (from both a) and b) channels).

d) A most significant objective, not yet satisfactorily achieved at this first stage of experimentation, appears to be a real-time mixing and integration between moments a), b) and c). The possibility to switch quickly and coherently from one computer use to another, and accordingly to really learn and understand at individual level by trial-and-error, stands in our opinion at the core of the cognitive revolution that a well planned use of simple computer technology could convey, at all learning levels [15].

2.2.3. Integration between different modes of analysis of the same physical situation. In particular we schematically characterize three *modes* whose integration, in our opinion, is necessary at any level of physics understanding:

a) Qualitative or phenomenological approach to the situation, whose salient features are firstly seen in terms of interactions among selected systems, and corresponding relations among selected variables [13]. It represents the crucial node where physical knowledge ("preformalized" in terms of topology, order relations, proportions, and "prestructured" in correlations and casual links) is referred to, and differentiated from general or common knowledge [7,9,10, 17].

b) Quantitative, "exact" formalization as a key understanding and handling of the simple, schematized experiment (as contrasted to the complex, articulated experience): again a crucial node allowing to explore and validate the correspondence (fit) between physical concepts, aspects of facts and mathematical relationships. (This *mode* is pratically the only one present in traditional courses, quite often without enough critical awareness).

c) Quantitative "approximate" handling of the complex, real phenomenon, where different aspects of theory, facts and formalism need to be adequately superposed-and-disentangled. Only this kind of process makes in fact clear the problematic meaning of the fit of physical theory to facts, and of matematics to physics [7,8]. Facing ***simple*** situations where the formalization can become non-linear, non univocal, non-solvable... and so on, approximation strategies (series expansion, numerical calculus, cuts in phase space, etc) and heuristic, looping approaches become essential to successful schematization.

On one side, it is worth to realize how much points 2.2.1 and 2.2.3 are tightly correlated to each other, while point 2.2.2 can play the role of instrumental hinge around which such a correlation can be articulated and enforced. As already remarked this is clearly a problem faced also by ITS research: however, on the basis of our experience, we believe that only through a cultural investment of this kind, stressing individual responsability in understanding, can the introduction of the computer significantly change our teaching-learning universes.

On the other side we'd like to point out that a pedagogical strategy based on the gradual reciprocal adjustment among several approaches, points of view, ways to look at... seems to be definitevely more efficient in the long term than a series of well defined *steps*. Provided, of course, that what is being done makes sense to the learner (and to the teacher), and that s/he is adequately supported in overcoming initial difficulties. This is at variance with the indications from some lines of research, (see for instance [18]), and in agreement with others, as for instance the ones represented by a quotation from "The Geneva Manifesto of Intelligent Learning Environments" [19]: <Multiplicity promotes articulation of knowledge: multiple situations for training transfer; multiple viewpoints because a constructivist approach requires that the learner be allowed to use and test its own relevant knowledge structures; multiple coaching strategies because none is universally efficient or adeguate. Learning requires frequent shifts between the various situations, viewpoints or strategies>.

3. The research

Two main features contribute to the novelty of this research, and to the applicability of its results to optmize physics teaching. They are:

i) The use of the computer, mainly but not exclusiveley in its on-line modes, not only as a teaching aid (i.e. dedicated to computation, exemplification, illustration, representation etc), but mainly as a cognitive instrument: contributing to the basic cognitive dynamics by which the structures of physics can become really acquired and assimilated.

From this point of view we underline the delicate and flexible blend between guidance (worksheets, homework, lectures, model-results etc) and openness (self adjusting of the conceptual paths) upon which rests the success of this kind of approach. Teaching efficacy, in fact, is linked to the very possibility of continous evocation, during the work, of previous ways to interpret the same context, and of continous addressing of the cognitive attention to selected, meaningful particulars. But stimulated evocation and addressed attention can very easily degenerate into the dull and uneffective forms of teaching we all know of.

ii) A cognitive awareness, developed in previous work and corroborated through the research, that privileges the strategic aspects of cognition: from the most naive ways-to-look-at to the most elaborated formalization criteria. It becomes in this way possible to single out the powerful links and "bridges" [20] which connect physicists' ways to interpret the world not to fragments or excerpts of spontaneous conceptions, but to the basic strategies and cathegories which subtend natural as well as scientific thinking. In other words: to look at a phenomenon in differential (time by time, equilibrium-like) ways, or in integral (across time, conservation-like) ways is a fundamental aspect of cognitive dynamics, implicit not only in any kind of science but in all everyday's interpretations of the world. It is therefore crucial that physics' teaching (i.e. equilibrium and conservation in mechanics) is able to invest, and to control, the particularizations and ambiguities by which such strategies are already adapted in any individual to their everchanging roles. (The same comment can be obviously made for a half-dozen of other cognitive strategies heavily involved in the first learning of physics). Still in other words: the deep "sources" of misconceptions are mostly the same upon which rest all the networks of efficient and unescapable commonsense's strategies, and all the constructions of formalized disciplines (physics at the first place). It is therefore necessary (and efficient) that models of thinking, models of discipline structuration and models of explaining/understanding, teaching/ learning interactions become more and more reciprocally adjusted and coherently articulated.

Open environment approach and integration represent then necessary facets of a research aiming to study the conditions by which natural thinking, physics theory and practice, and teaching-learning paths can be driven into reciprocal resonance and to sound cultural outcomes.

4. The intervention

In the academic year 86-87, at Naples University, we have started a long term research project aimed at investiganting understanding problems and productive teaching approaches in UIPC for physics majors. Such courses are traditionally located in Italy at first and second year (I: mechanics and thermodynamics; II: electromagnetism and optics); two other parallel independent courses take care of the laboratory activities. On the structural ground, the short term goal of the project is the integration to the current courses' structure of computer-supported activities, to allow for qualitative changes in the teaching-learning interaction. Our long term goal is a completely new structure for the different courses, with a deeper correlation between theoretical and laboratory ones.

We faced, as usual, several difficulties: partly related to the task, partly rooted in resistences, in students and in the academic community, toward non trivial changes in familiar UIPC patterns.

4.1. The teaching context

In the following we refer about a first-approach intervention in the mechanics-thermodynamics theoretical courses normally structured in lectures (about 70 h) and problem solving classes (about 70 h). About 110 students have been involved in the last year (one of us was one of the two UIPC teachers).

The teaching experiment has been articulated in two phases: an initial activity (about 20 hours) at the beginning of the academic year, compulsory for all students; and a second one along the IPC, on voluntary basis (about 6 hours plus homework). In the first phase the students have been exposed to a combination of lectures and demostrations with on-line ultrasonic motion detector (4h, groups of about 60 students), followed by correlated computer-based laboratory activities (5h, 3 students' groups); and to introductory problem solving classes (10h, about 15 students' group) on elementary physics subjects, including one-dimension uniform motions and floating-sinking situations. In the second phase students (in groups of 3) have studied motion's kinematical and dynamical aspects and some thermic phenomena, always on the basis of on-line data. (cfr. § 4.1 and 4.2)

In the research work the main interest was focused on physics, to integrate more phenomenological awareness and more understanding into the rather abstract IPC content to be learned. Therefore, within resources' and structures' limitations, we have choosen to propose essentially OEA in laboratory activities, with user's frendly MBL apparatus [21] which require no informatic prerequisites; - to use some knowledge-based software for simulations; not to propose, in the initial experimentation, student's programming activity; to stress, from both points of view of innovation and of research on innovation, the ways by which meaningful and flexible understanding can be achieved when the cognitive path can be assisted by the use of informatic technologies.

It is out of the limits and the scope of the present paper to exhaustively account for the experimental trial of all objectives and hypotheses underlying the intervention, and for the further problems raised by such a work. We will therefore restrict ourselves to a short description of a specific example about the study of motion and a list of the activities which have been tested in the last years, with comments on some typical aspects of students' work.

4.2. An example of integrated activities in the study of motion

The students started from kinematics of body's motion, studied by means of the ultrasonic motion sensor. This activity was aimed at linking the direct interference among natural perception, motricity, language and experience with the formalization of complex variables such as velocity and acceleration, and of spaces of variables. The laboratory work was supported by long worksheets with hints and open ended questions. The students worked in groups of three, with about a tutor every three groups.

Fig. 1 shows typical on-line records obtained when a student is walking in front of the motion sensor according to the (deliberately vague) instruction <Move back and forth, as regularly as you can, in front of the motion sensor. Discuss the on-line graph of your movement>. The problematic correspondence between space and velocity plots (the last one shows very clearly the effect of single walking steps) was always a starting point for fruitful comments about the correlation between (x,t) and (v,t); and the question <What is going to happen with acceleration?> followed quite naturally. From records of this kind (and from comparison with numerical tables of data) it is also possible to analyze and discriminate physical sources of irregularities versus intrumental and noise errors. This kind of work helped in dealing with some a priori common, instruction-resistent difficulties, as the following ones: - confusion between the geometrical trajectory of the motion and x(t); - the interpetration of an x(t) graph as it was a v(t) one and viceversa; - correlation of an inversion in motion with a change in the slope sign of v(t) instead of a crossing of the time axis; - in general, difficulty to allocate features of graphical representations in terms of physical characteristics and viceversa; - meaning of negative velocity and negative acceleration; - in general, difficulty to correlate features of different variables-behaviours representing the same motion ((x(t) vs v(t) vs a(t), etc...) .

Parallel to this laboratory activity, in the exercises' classes, it has been proposed: - to describe in words how to realize the motion represented by the given graph (and viceversa); - to read data from hard copies of on-line x(t) graphs of real motions and calculate numerically by hand v(t) (similarly from v(t) to a(t)); - to compare the obtained results with the corresponding computer graphs, discussing the different results. This task, based on operational use of data from real motions, allowed also meaningful contacts with simple topics related with numerical calculations, approximations, smoothing of curves, etc...

The representation aspects, in their general form, were also addressed by operations with linear functions in different abstract spaces: weight versus volume plots, to describe complex floating-sinking problems, compared to distance versus time in uniform straight

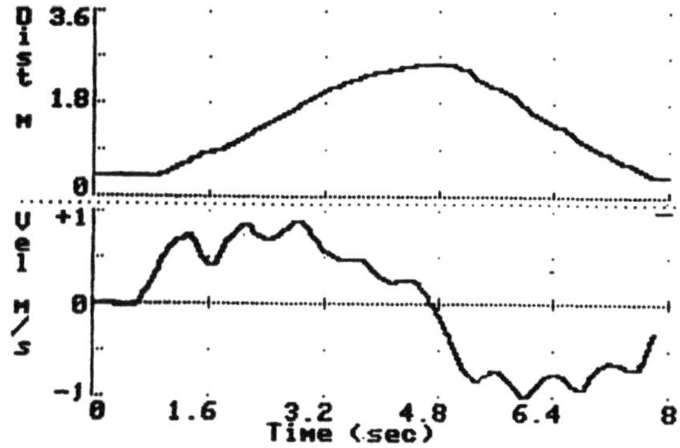

Fig 1

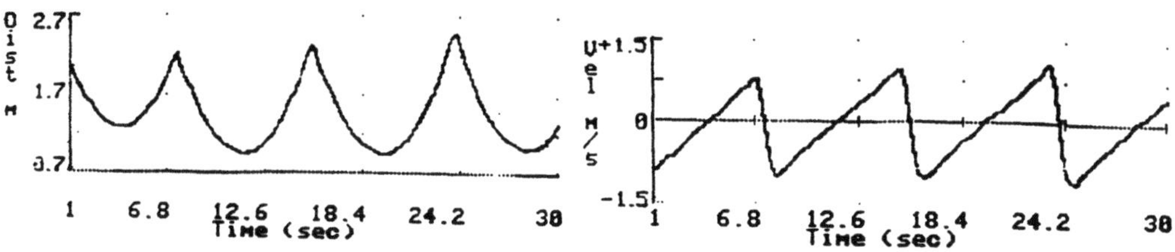

Fig 2

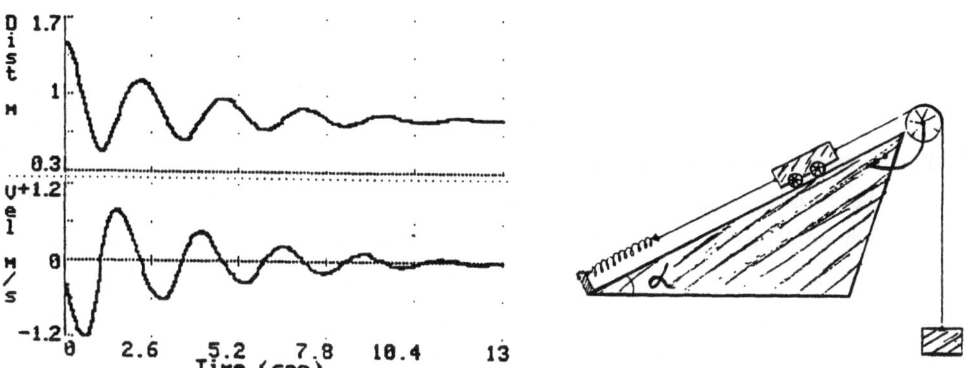

Fig 3

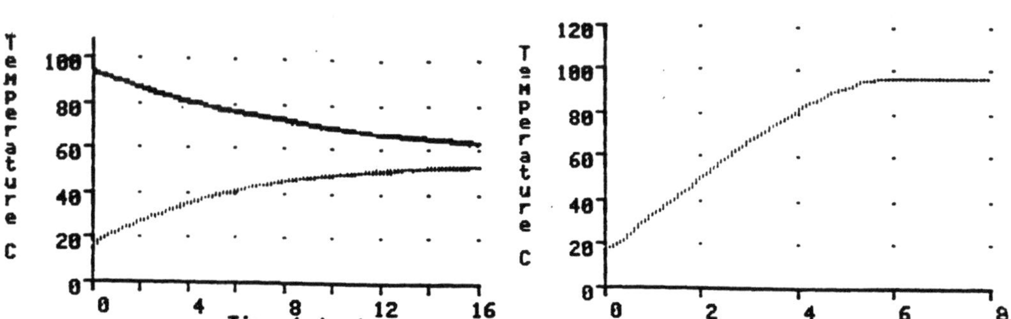

(a) Fig 4 (b)

For captions see text

motions. Learning how to deal with different meanings of the same "forms" or trends was helped by being able building quickly many quasi uniform straight motions with the motion sensor.

All students, about 110, have been exposed to 3 questionnaires at 3 stages of the work (initial, intermediate, final), aimed at testing the capabilities of interpetring and predicting kinematical graphs. The analysis of these data is in progress; the preliminary results indicate a substantial improvement, during and after the MBL activities, in the above mentioned capabilities. Transversal effects of this work have been seen in several other tasks during the course, for instance when correlations of kinematical graphs for a generic motion problem were needed in mechanic experiments and exercises.

A second type of activities was related to the study of a spring and mass system. In the usual Laboratory course the students measure the elastic constant of quasi massless springs both statically and dinamically via oscillations' period measurements, submitting written papers in which is requested a detailed error analysis. Some groups of voluntary students (about 10%) have integrated such compulsory lab tasks by studying the spring/mass system by means of the motion sensor. Firstly the same quantities were measured for quasi massless and massive springs and compared; this allowed a first study of x(t), v(t), and a(t) for the oscillations and a discussion about the two different error environments.

At the same time the simultaneous use of a force sensor to which the spring was attached and of the motion sensor looking at the suspended mass allowed to experiment clearly the correlation between force and acceleration in the harmonic motion and a first reflection about dynamical aspects of motion.

Later these students experimented with damped oscillations difficult to measure without the on-line data collection. In this way more complex phenomenology was approached by integrating, in the laboratory setting, the MBL activities with the use of a specially designed simulation software [22], for studying one damped and/or forced unidimensional single oscillator and two, three, seven coupled oscillators. This integration allowed to verify hypotheses and to improve the analysis method. Some experimental study of the influence of shape, fluid resistance and mass on the damping process was therefore possible. The simulations allowed: to play with parameters' variations; to interpret the data by models in accordance with theory; to explore situations not reproducible in laboratory experiments. When the simulations' predictions were not in agreement with the MBL data, the need of taking into account other effects was reinforced; often meaningful "divergent" cognitive dynamics set on, which had to be stopped to mantain a reasonably effective teaching scheme (cfr. § 2).

While the design of the intervention was based on the above described "integration and open enviroment approach framework", the detailed sequences and schedules for specific activities came out from the process through a dynamic adaptation of the design to the speficic contexts.

4.2. Other activities

In the following we list activities which have been tested in the last years. The point a) is included in the more detailed description of the previous section.

a) Kinematics of body's motion.

b) Inclined plane.

(Kinematics and dynamics, in different friction and slope conditions).

Fig. 2 shows typical on-line records obtained when a rather heavy ball is periodically climbing up and rolling down an inclined plane according to the instruction < Let the ball move up and down the inclined plane as regularly as you can, by briefly stopping-pushing it at the bottom of the incline>. Note the effort needed to adjust manual intervention to physical patterns and to discriminate and integrate the roles of the physical law on one side, of the initial and boundary conditions on the other one.

c) Springs.

(From quasi-linear, quasi-conservative, quasi-massless...behaviour to non linearity, energy-dissipation, self-weight effects...; both Motion and Force Sensors have been used).

d) Oscillating roller-skate.

(See sketch in Fig. 3. The system combines different sources of weight, inertia, friction, self-dissipating elasticity... etc, all with adjustable parameters to allow for a disentangling of different contributions; both Motion and Force Sensors have been used).

Fig. 3 shows typical on-line records obtained by the motion detector coupled to the oscillating roller skate. Note that motion is not exactly periodic, damping is not exactly exponential, etc.... By a patient analysis one becomes gradually aware of the basic character of *constraint* of any physical law, enforced by relations among variables shaping the available phase space (approximate vs exact solutions of motion equations, comparison with conservations' laws, etc..).

e) Three-sources (2 variable systems + environment) heat exchanges.

(From different ways to evidence specific heat - mixing, contact, heating - to the study of variables, parameters and boundary conditions in heat conduction. Two temperature probes together with traditional thermometers have been used).

Fig. 4 shows typical on-line records from: A) temperature detectors coupled to two different quantities of water exchanging heat through a thin wall, and rather well isolated from the environment; B) temperature detector coupled to water (not purposefully isolated) heated by a constant power.

5. Conclusions

We would conclude from our experience that the educational use of computers in teaching-learning physics can very powerfully support all levels of the cognitive path (even at different school levels).

However we underline the necessity that the cognitive path itself, even when assisted and addressed by the use of computers, preserves two crucial characters: i) continuos reference to the total of the experience, knowledge and culture directly or potentially involved by the learners in their cognitive paths: when this reference is uncorrectly or unexplicitily handled, then *misconceptions* and *misinterpretations* are reinforced or produced; ii) tight correlation with the specificity of the real situation, in its material, experimental, instrumental, formal and cognitive aspects. From this points of view, simulations (of both physical and mental behaviours) can be very helpful for students, teachers and researchers; however we cannot see, by now, how they could substitute an intelligent, human guidance throughout the complexities of the learning process.

References

1. International Journal of Science education, special issue: Student Conceptions in Science (R. Driver editor) vol.11, n.5, 1989
2. Sassi, E.: The computer in Sciences' Education (in Italian). In: Le Nuove Tecnologie nei processi formativi - Informatica e Telematica, Studi e documenti degli Annali Pubblica Istruzione, n.41-42, pp.71-89 Firenze: Le Monnier, 1988
3. Sassi, E.: Using educational software in schools: features, possible roles in curricula, integration problems (in Italian). In: Computer e software didattico (G. Marucci editor), Lisciani e Giunti 1989
4. Balzano, E.; Sassi, E.: TRASFORM: An Open Environment to Study Affine Geometric Transformation on the Plane. Educational Training Technology International, 27, 1, pp.92-109, 1990
5. Olimpo, G.; Sassi ,E.: Notes on researches about educational uses of computers in physics education (in Italian) La Fisica nella Scuola, special issue on "L'elaboratore nella didattica della fisica", vol.XXIII, n. 2, pp.6-18, 1990
6. Sassi, E.: Integrating Phisics and Computer Education in a Single Process in Designing Computer-Based Learning Materials (H.Weinstock and A. Bork editors) Nato ASI series, pp.183-207, Berlin-Heidelberg-NewYork: Springer 1986
7. Moretti, M.: Difficulties in learning classical mechanics by physics majors, in Introductory Physics Course I (in Italian.) Degree thesis, prof. P. Guidoni, E. Sassi , Naples University, Department of Physics, 1988
8. Sgueglia, G.: Learning difficulties in Thermology: computer based experiences and problems for Introductory Physics Course for physics majors (in Italian) Degree thesis, prof. P. Guidoni, E. Sassi , Naples University, Department of Physics, 1989
9. Guidoni,P.: A phenomenological Approach to Energy Ideas. In ;Energy Matters (R. Driver and R. Millar eds) pp.103-132. University of Leeds, Fairbain House 1985
10. Arcà, M.; Guidoni, P.: A Phenomenological Approach to Mechanics (In Italian and Spanish). In: Proceedings of IV Curso Internacional sobre Ensenanza da la Fisica. State University of Bogotà, 1987
11. Arcà, M.; Guidoni, P.; Mazzoli,P.: Structures of understanding of the root of Science Education. European Journal of Science Education.Part I : Vol 5, 1983, pp.367-375; Part II : Vol 6, 1984, pp.311-319
12. Guidoni, P.: On Natural Thinking. European Journal of Science Education. Vol. 7, 1985, pp.133-140.
13. Arcà, M. ;Guidoni, P.:Looking by Systems, Looking by variables (In Italian). Torino, EMME Edizioni, 1987
14. Guidoni, P.: Contrat Didactique et Connaissance. In: Actes de la IIIème Journée Internationale d'Etude sur Médiation et Remédition Didactique. Université de Marseille, 1988

15. Guidoni, P.: The Computer as a Cognitive Instrument (in Italian). In: Il Piano Nazionale per l'Informatica, pp. 94-103. Napoli, Loffredo Editore 1990
16. Guidoni,P.: But does it exist a Physical Thinking ? (In Italian). Presentation to: J.Piaget: The Physical Thinking (Italian version). Milano, EMME Edizioni, 1985
17. Arcà, M.; Gagliardi, M; Guidoni, P.; Mazzoli,P.: Observations about the Physical, mathematical and cognitive foundations of Mechanics (In Italian). In: Proceedings of the Ist National Meeting of GNDF, pp14/1-14/36. University of Palermo, 1982
18. Scanlon, E.; O'Shea, T.: Cognitive Economy in Physics Reasoning: Implications for Designing Instructional Materials
19. Mendelsohn P. et al: The Geneva Manifesto of Intelligent Learning Environments, TECFA Document 90-2, Faculté de Psycologie et des Sciences de l'Education, Université de Genève, Suisse
20. Clement, J.; Brown,D.E.; Zietsman,A.: Not all preconceptions are misconceptions:finding "anchoring conceptions" for grounding instruction on students' intuition, in International Journal of Science education, special issue: Student Conceptions in Science (R. Driver editor) vol.11, n.5, 1989, pg 554 - 565
21. The ultrasonic motion detector was first developed at TERC, 2067 Massachussets Av. Cambridge, MA 02140, USA; an improved version and the force and temperature probe has been developed by the project "Tools for Scientific Thinking" at the Center for Science and Mathematics Teaching (director R. K. Thornton), Tufts University, Medford, MA 02144, USA
22. Balzano, E.; Cerbarano, F.; Cianciulli, E.; Sassi, E.; Sgueglia, G.: Oscillators and Oscillations (in Italian), Progetto Strategico T.I.D/C.N.R.; software and manual distributed by Società Italiana di Fisica, 1990

Practical Work Aid: Knowledge Representation in a Model Based AI System

Joël Courtois

Institut Supérieur d'Electronique de Paris, 21 rue d'Assas, F-75270 Paris Cedex 06, France, and
LAFORIA, Université Paris VI, 4 place Jussieu, F-75252 Paris Cedex 05, France

Abstract: SIAM is a model-based expert system designed for diagnosis aid with training in mind. SIAM is meant for students in higher education (polytechnic or university level) confronted with technical problems during practical work. Its aim, on the one hand, is to help the student on the spot, as quickly as possible and, on the other hand, to make him rapidly autonomous in his operations by teaching him a diagnosis method. The system is adaptable to various kinds of laboratory equipment in different fields of application: electronics, optics and mechanics. In this context, designing an intelligent tutoring system implies providing advanced technical and pedagogical abilities. Several levels of metaknowledge are required, which enable it, on the one hand, to be as independent as possible of its fields of application, and on the other hand, to describe in a pedagogical and relevant way, the key steps of the cognitive process followed by the experts. Most of the knowledge is represented by using models and descriptives. The advantages of this approach are both technical and pedagogical, by allowing easier acquisition of knowledge, reusability of the models and construction of relevant explanations about the diagnosis methodology. Maintenance aspects have also been considered so that the evolution of the system can be directly monitored by the teachers. A series of experiments using SIAM enabled us to measure its impact on the students and to evaluate our design choices.

Keywords: knowledge representation, model-based diagnosis, metaknowledge, genericity, maintainability, ITS

1. Introduction

Knowledge based systems are often long and costly to design and construct. They usually work in very restricted areas where, it should be said, they perform quite efficiently. The use of these artificial experts in a training course, poses at least two problems :

NATO ASI Series, Vol. F 86
Intelligent Learning Environments
and Knowledge Acquisition in Physics
Edited by A. Tiberghien and H. Mandl

- like human experts, will they have the indispensable pedagogical capabilities for the teaching context ?

- how much time (and money) will be necessary to create all the expert systems corresponding to the numerous chapters of a mathematics, a physics or an electronics program ?

To try to answer these questions, it was decided to carry out an experiment, in a practical domain, whose prime objective was to arrive at a system that could be implemented. We therefore used a technique of artificial intelligence expert systems, as a starting point for formalizing technical knowledge. Then we approached the problem step by step in a very pragmatic way, getting the students to participate as early as possible. In this way, the first experiments were taken advantage of to improve the system.

The SIAM system which will now be presented is the result of this research work. SIAM is a pedagogical system designed to supply students with the right methods of diagnosis. It is also able to adapt itself to many technical fields of application. SIAM uses several levels of metaknowledge which enable it both to be as independent as possible of its fields of application, and to describe in a pedagogical and relevant way, the key stages of the cognitive process followed by the experts.

2. Objectives of the System

2.1 Functional Specifications

The domain chosen for experimentation is that of tutorials in higher education. This context is interesting from several points of view.

Firstly, it is necessary to draw on expertise in a large number of domains (physics, electronics, optics, etc.) and in several different types of materials.

Secondly, one must have the pedagogical skill needed to make the students autonomous as soon as possible. Lastly, it must be designed within a reasonable time because of the frequent modifications of the tutorial subjects linked to the evolution of the programs.

The use of an Intelligent Tutoring System is also a useful complement to the work of the human tutors, for correct learning, in a practical context, needs very close surveillance to be efficient in a short time.

What are the components of this tutorial assistance and what can we expect to simulate with an ITS ?

Human tutors have a fundamental role in the acquisition of the reasoning linked to concrete applications. This makes the student progress from the stage of beginner to that of expert while becoming aware of the links that exist between theory and practice.

At the highest level, the tutor may have to reiterate part of the theory so as to specify the context and the methods that can be used.

The theoretical aspects are also approached from a more pragmatic angle to explain why the application being studied, works or does not work.

Finally, practical assistance takes up most of the time since its aim is to solve the numerous problems that crop up during the manipulations.

It was for this last point that the SIAM system (Intelligent System for Manipulation Assistance) was designed.

Let us now see what this practical assistance entails. It is relatively complex and we can distinguish three principal sources of problems:

- a misunderstanding of the work expected ; this can come from incomplete knowledge of the underlying theory or of one of the materials involved ;

- an error in the carrying out of the manipulation ; this is the most frequent type of problem ranging from an assembly error to confusion about materials or to any possible mistaken use ;

- a total breakdown or malfunctioning of a material ; this is also quite frequent because of the bad handling to which the materials are frequently submitted.

Furthermore, it is important to remember a major aspect of the tutor's function, that of pedagogical skill. Indeed, it is essential not to solve the problems for the student but rather to give them the knowledge and the know-how that will make them autonomous in a short time.

2.2 Defining and Formalizing Constraints

First of all, the operational goal that we fixed was to make the system work on several dozen tutorial subjects in domains apparently very different from each other (Electronics, Optics, Electrotechnology, etc.).

Each tutorial being made up of a set of minor manipulations, several hundred knowledge bases would have to be acquired both technically and pedagogically with the time constraints that can be imagined.

Therefore, it is imperative that the system should operate as if these knowledge bases were defined, even though it uses much more general knowledge of the domain and rather simple descriptions of the components involved in the various specific contexts. The following constraints appear:

- The necessity for the system to be able to solve the problem with which it is confronted ; this point requires the creation of a real expert in technical diagnosis ;

- The possibility of using the system for numerous tutorials in the same domain but also the possibility of changing domain ; the knowledge supplied must be of a high enough level to allow adaptation to several contexts ;

- A pedagogical approach in the search for solutions ; the system must be able to explain itself and to react in an interactive way with the learner. It is not a question of solving the problem but rather of guiding the student towards the solution by the presentation of the methodology of research followed by the "expert teacher".

A large number of choices of conception were made to respond to these three points whose most important ones are the followings.

The first point is that priority was given to the definition of descriptives rather than to the writing of rules in order to dissociate the context specific data from the procedures of data processing. The justifications of this approach must be found in the genericity of the technical diagnosis and the facilities of maintenance.

The technical expertise was focused on the individual diagnosis of the materials. This choice could be discussed but we must not forget that we find ourselves in prepared sessions "that should work" and so we know the principal interactions between the materials. These interactions are therefore introduced as possible malfunctions at the individual level.

Another important choice, when the system was devised, was to give not too prevailing a place to the student's model. As a matter of fact, we must not undertake a follow-up over a long period but rather operate on a one by one basis with a large number of users. Nevertheless, as the system must be used by students of different classes, an adaptability to the user's level enables the system to guide the diagnosis and to determine the level and the quantity of the explanations to be supplied.

Userfriendliness is obtained by an interface in natural language, an ATN [24], keywords menus and "questions and answers".

Finally the system was not devised to take the place of the human assistant but rather to work as a computerized complement.

3. SIAM Architecture

3.1 General Presentation

The general architecture of the SIAM system is proposed in the figure of the next page. It is based on the use of Models (a) and Descriptives (b) conceived from the models. The needed descriptives collected together define the Context of Work (c) for the system.

SIAM technical knowledge can be divided into two principal categories: on the one hand, the descriptive knowledge, and on the other hand, the expert modules. Most of the system knowledge is represented in a descriptive way. The Production Rules (d), in the expert modules, are very general ones able to work with any of the descriptives made with the predefined models.

The Inference Engine (e) has no specific knowledge. It only actives rules and metarules of the knowledge base.

Interfaces (f) allow understanding and generation of natural language for communications between the Student and SIAM. The Menu technique is also used by the system.

3.2 Descriptive Knowledge

Descriptive knowledge given to the system through descriptives contain different types of information :

- conceptual descriptions ;
- functional and relational descriptions ;
- physical descriptions.

The first category supplies the most general knowledge, the highest abstraction level which the expert's behaviour obeys. The extraction of these cognitive processes is often difficult because it comes up against the expert's judgement "it's obvious" or "it's logical".

It is, nevertheless, at that level that we find the most useful knowledge, both to obtain a greater efficiency of the diagnosis system and a greater relevance of the pedagogical method being implemented.

For example, we may find the conceptual description of a tutorial (what a student is going to do in a tutorial) and the conceptual description of a measure and a signal (we also had to describe the different types of signals which could be studied - light, electricity, motion).

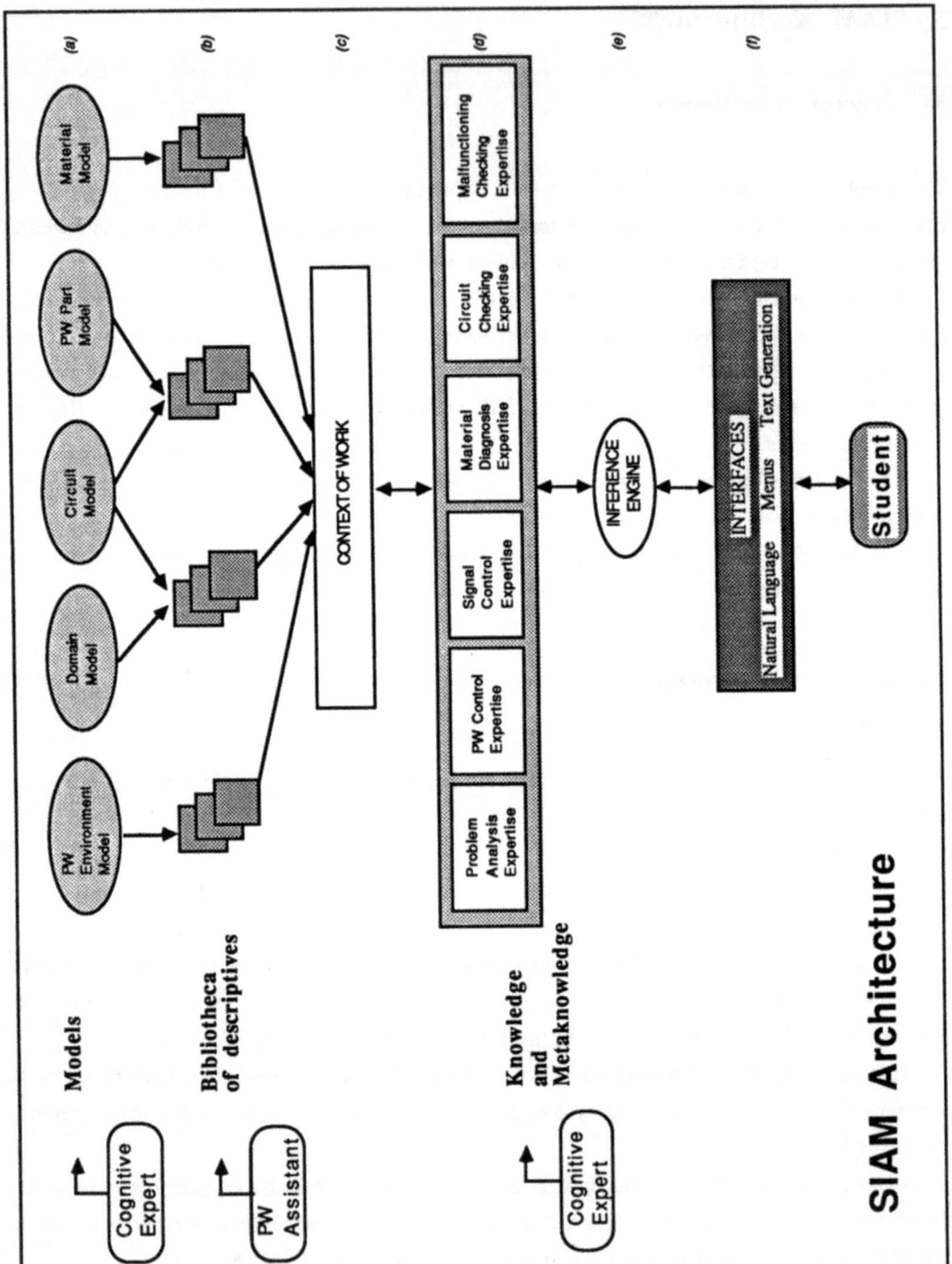

Fig. 1. General architecture of the SIAM system

The second category allows us to determine the function of the different elements which are used in the manipulations, as well as the main possible interactions due to the propagation of other elements' problems.

Here the whole technical knowledge of the expert is thoroughly involved.

In this category, we find the functional and relational descriptions of the devices. These descriptions of the various materials are made within the same frame ; this descriptive constraint has the advantage of making it possible to write diagnosis rules quite dissociated from the materials and their application fields. Up to now, this structure has enabled us to describe all types of materials from the oscilloscope to the collimator, as well as the lens and the simple wire.

The third category, the easiest to carry out, enables us to represent the various elements and their kinds of connections physically. Here we find the physical description of the materials (so as to detect the risks of confusion between some of them), the physical and functional description of the circuits (a list of the materials and the links between them in the circuit as well as the way the circuit is carried out and the way it can be controlled) and the description of the precise context of a tutorial problem.

The models which define the structure of representation of this knowledge are based on frames and are the following ones:

-Practical Work Environment Model: allows the representation of the objectives of practical work sessions. The principal attributes are PW_Success_Conditions, Measurement_Success_Conditions, Signal_Success_Conditions, Problem_Description.

- Domain Model: used for information related to the domain, the signal description in particular. As for the signal description, we have attributes like Name, Detection_Complexity, Measurement_Means, Origin, Transit.

- Material Model: used for all kinds of material from all domains. The most important attributes (see next figure) are Role, Directions_for_Use, Setting, Non_Setting, Consequence, Test, Symptom, Limit, Non_Respect, Physical_description.

- Circuit Model: this model describes from the one hand, the circuit physically and from the other hand, its strategy of checking which is specific to the domain of application. No descriptives are made through this model, the corresponding information being included into the domain descriptives for the strategy of checking, and into the practical work part descriptives for the physical description.

- Practical Work Part Model: allows the description of each of the minor manipulations which constitute a session (e.g., materials, circuit, actions) ;

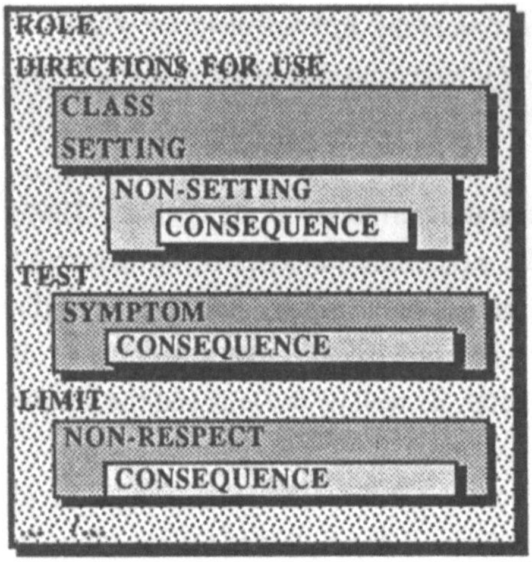

Fig. 2. Part of the Material Model

Bibliotheca of descriptives were conceived through those models on 3 domains of application, 50 materials and 40 parts of practical work.

A specific context of work for SIAM is made by gathering of the required descriptives. With all these data, the system just needs the student's problem description to be able to begin its investigations thanks to the expert modules.

3.3 Expert Modules

The descriptive knowledge is used by a set of general expert modules thoroughly independent from the context, the fields of application and the materials.

These knowledge bases are managed by a large quantity of metaknowledge integrated into the various modules. On the one hand, the metaknowledge must control the local reasoning within the modules and, on the other hand, must provide the control transfer between the various modules. The metaknowledge levels are not detailed and do not exist in a set way ; the system itself chooses, at any given time, to which level the control is transmitted dynamically and without any pre-existing order.

The principal expert modules are the following ones:

- Tutorial Control Expertise ; this high level expertise (PW Supervision in the next figure) tries to verify if each of the necessary conditions for a correct carrying out of the practical work is present. It principally uses the conceptual descriptions. From this level, metaknowledge can transfer the control to the theory control expert module or to the calculus control expert module or to the measurement control expert module ;

- Problem Analysis Expertise ; from information supplied by the student, this expertise tries to deduce the largest possible quantity of data on the real state of the manipulation. From this level, the control is transmitted to the tutorial control expertise ;

- Signal Control Expertise ; this high level expertise will be activated when the functioning of the studied signal seems to be in doubt. From this level, metaknowledge can guide investigations according to three axes:

. signal existence problem ;

. signal state problem ;

. signal value problem.

- Material Diagnosis Expertise ; this works following several steps directed by metaknowledge: finding of the whole set of defective settings, organization of these causes according to their complexity, dialogue with the student to make checks, troubleshooter tests, control of the borderline cases of use. The main rule applied in this expertise is :

KNOWING: the problem is X;
there exists a device A,
whose *Consequence*
of the *Non-Setting*
of a *Setting*
of the *Directions-for-use*
is X;
THEN: The device A is open to be the source of the problem.

- Circuit Checking Expertise ; this expertise will enable the student to examine and thus to control his circuit. The way of control to be followed is explicitely found, step by step, in the domain description thanks to the circuit description attributes.

- Malfunctioning Checking Expertise: the order of control of the devices is deduced from their rate of breakdowns and from their difficulty of testing.

Some of these modules contain, themselves, several levels of metaknowledge. The various expert modules must operate as if they were autonomous agents able to transmit the control to each other, each time it is necessary.

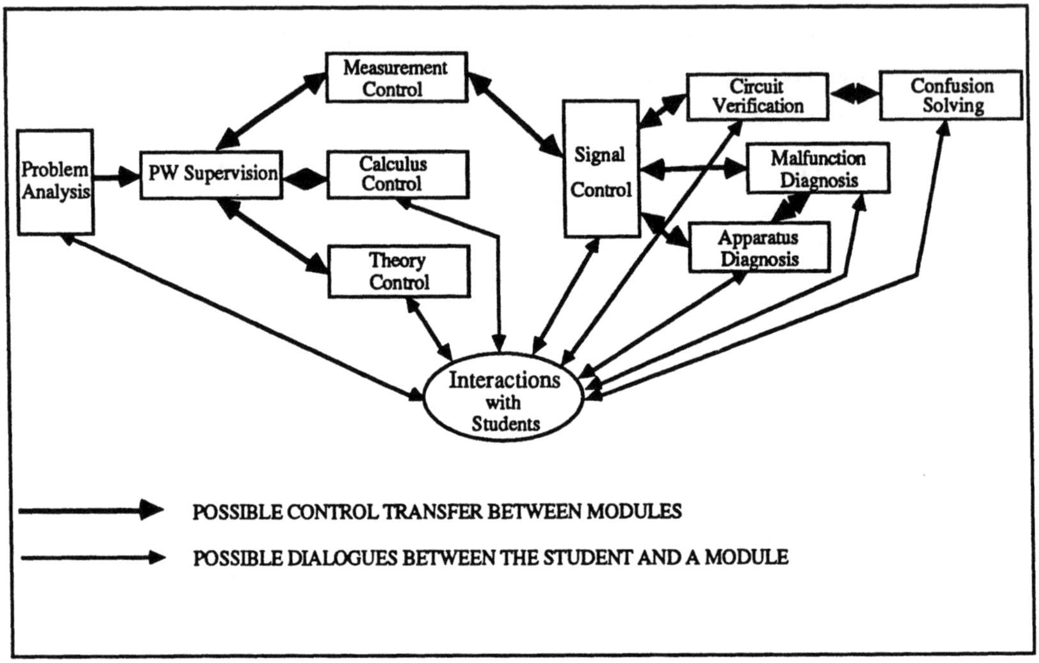

Fig. 3. Examples of control transfer and dialogues with the student

Examples of possible control transfer between the modules with the indication of dialogues with the student are given on this figure.

The whole system's knowledge, descriptions and expertise, is stocked within a single format close to SNARK's [13], that is to say, a production rule based system in predicate logic. This single representation enables the system to handle, in the same way, independent facts, structured descriptions, rules and metarules.

3.4 Pedagogical aspects

Our aim is to guide the student during diagnosis sessions by giving him all the necessary justifications concerning the method followed by the expert and all useful technical explanations about the materials involved.

The practical use, in a technical context, is the best way to help the student to understand the theoretical principles learned in a classroom. The SIAM system does not teach diagnosis methods explicitly. The student acquires the expert methodology thanks to the example the system gives during the diagnosis sessions. The quality of the SIAM comments is then a very important point. It has been obtained thanks to the approach chosen for the designing of SIAM.

Pretty interfaces are necessary to give more userfriendliness in the relations between the student and the system.

The first contact is realised with a natural language interface that allows the student to explain the problem which occurs. In order to avoid a situation of misunderstanding, if some trouble appear with this interface, another one based on the technique of keywords and menus is used as a back-up.

During the session, the explanations given by the system and the questions asked to the student are dynamically conceived thanks to a specific interface that uses predefined pieces of sentences linked to the objects manipulated by the system: rules and attributes. The general structure of the sentences is defined with some special actions contained in the rules. The specificities of a context are found in the messages linked to the descriptions of the domain, circuit, materials, etc. The possible answers to the questions asked by the system are given in a menu which is also dynamically conceived.

At the end of a session, the student can give some comments wich contain technical indications or appreciations concerning the SIAM system behaviour.

4. SIAM Evaluation

The system presented above has been evaluated during one year (1988-1989) in one of the laboratories of the Institut Supérieur d'Electronique de Paris, where students can use it in three domains, optics, electronics and electrotechnology. This real experimentation has been very valuable for the achievement of the system.

The students' reactions have given us a large amount of information. There is no psychological problem relative to the acceptance of the system by students. It is just necessary for them to understand exactly what the field of application of SIAM is.

The way the system works is satisfactory and corresponds well enough to the expert's behaviour. The student is able to follow the whole cognitive process followed by the expert which enables him to understand his future problems better.

It is important to notice that the student is normally seldom supplied with this kind of information in tutorials, because the assistant usually explains why it did not work once the solution is found out, but does not systematically specify how he has detected the origin of the problem (whence the experts' magical aspect).

The experts can not always be reproached with the lack of explanations of this "how", for they very often do not know at all how they manage themselves. As a matter of fact, experts work by automatisms. The knowledge they use is *compiled* and is no longer accessible.

The amount of explanations which do not depend on any artificial model of the student, must be precisely adapted so that the students should not consider the system as too talkative or too fast. During the system development, it was necessary to condense some descriptions which appeared to be too detailed or, on the contrary, some others had to be more detailed, so that the precise origin of the problem could be found faster.

The natural language interface, which is not detailed herein [10], must contain no errors because users would get tired of it very quickly. A graphic interface could improve the userfriendliness even more for the circuit checking.

Finally the system fulfils its last aim which is to create new tutorial subjects rapidly. As a matter of fact, modification or creation of a practical work subject, pedagogical aspects included, can be done and tested in a very short time: if all the devices needed are already described, it will take about one hour of work to describe a new part of a practical work session.

5. Conclusion

In this research work, the adopted point of view consisted in giving the maximum of explicite knowledge to the system, so that it could control the development of its actions itself, and that thanks to a large quantity of metaknowledge. The model-based representation of the most important part of this knowledge and the conception of general production rules are the principal characteristics of this work. They both offer the opportunity of a reusability of knowledge and a facility of maintenance that allows the teachers to monitore themselves the evolutions of the system.

The obtained results have justified these choices both in the system's efficiency, possible evolutions and adaptations, and also pedagogically speaking.

However, to provide the SIAM system with a pedagogical approach, it was necessary to devise a thorough analysis of the experts' motivations when investigating.

The system does not use a specific pedagogical module, since the explanations are constructed while using the expert's knowledge, by concatenation of predetermined bits of sentences linked to the used objects. These sentences were put together both by the technical expert and the pedagogue.

Once that task is done, SIAM diagnosis can perform for all to see and then, the student discovers all the expert's cognitive processes. It enables the student to understand better how the assistant detects the origins of the problems.

This point was fairly well understood by the human assistants themselves who had to reconsider their own methods of arriving at an explanation. As in most domains, the extraction of the knowledge of a human expert leads him to consider his methods of work a new point of view.

From this case, SIAM has proved some of the advantages of ITS and the improved system has demonstrated the great attraction of intelligent systems that know why they do what they do.

References

1. Anderson, J.R.: The Expert Module. Foundations of Intelligent Tutoring Systems, p. 21-53. LEA. Hillsdale. New Jersey. 1988
2. Brown, J.S., Burton, R.R.: Multiple representations of Knowledge for tutorial reasoning (SOPHIE). Representation and understanding. Academic Press.New York. 1975
3. Brown, J.S., Burton, R.R., De Kleer, J.:.Pedagogical, natural language and knowledge engineering techniques in SOPHIE I, II, III. Xerox Palo Alto Research Center. 1981
4. Burton, R.R.: The Environment Module of Intelligent Tutoring Systems. Foundations of Intelligent Tutoring Systems, p. 109-142. LEA. Hillsdale. New Jersey. 1988
5. Clancey, W.J.: Methodology for Building an Intelligent Tutoring System. Method and Tactics in Cognitive Science. Kintsch, Miller & Polson Editors. Hillsdale. 1984
6. Clancey, W.J.: From Guidon to Neomycin and Heracles in Twenty Short Lessons. ONR Final Report 1979-1985. Stanford Knowledge Systems Laboratory. 1986
7. Courtois, J.:.Système intelligent pour l'apprentissage du diagnostic technique. Congrès ITS 88. Montréal. 1988
8. Courtois, J.: Les apports de l'informatique dans la formation du raisonnement scientifique appliqué. Colloque "Les finalités de l'enseignement scientifique en question". Marseille. 1989
9. Courtois, J.: Gestion d'expertises multiples dans un système de diagnostic - Le système SIAM. 9e Journées Internationales d'Avignon. 1989
10. Courtois, J.: SIAM : un système de diagnostic qui s'adapte aisément à de nouveaux domaines et qui enseigne sa méthode. Thèse de Doctorat de l'Université Paris VI. 1990

11. Dague, P., Devès, P., Raiman, O.: Troubleshooting : when modeling is the trouble. AAAI 1987. Seattle. 1987
12. De Kleer, J., Williams, B.C.: Diagnosis with Behavioral Modes. 11th IJCAI 89, p. 1324-1330. 1989
13. Laurière, J.L.: Intelligence Artificielle (tome2) Représentation des connaissances. Eyrolles. Paris. 1988
14. Marrakchi, M.: Représentation des connaissances pour l'aide au diagnostic industriel: application au système expert S.E.DIAG. Thèse de Docteur Ingénieur. Université de Valenciennes. 1986
15. Moustafiadès, J.: Formation au diagnostic technique : l'apport de l'intelligence artificielle. Masson. Paris. 1990
16. Nicaud, J.F., Vivet, M.: Les tuteurs intelligents : réalisations et tendances de recherches. TSI vol.7, no.1. 1988
17. Paliès, O.: Métaconnaissance pour la modélisation de l'élève. Contribution au diagnostic cognitif par système expert. Thèse de Doctorat d'Université. Université Paris VI. 1988
18. Pitrat, J.: Métaconnaissance, Futur de l'Intelligence Artificielle. Hermès. Paris. 1990
19. Reiter, R.: A Theory of Diagnosis from First Principles. Artificial Intelligence 32, p. 57-95. 1987
20. Raiman, O.: Order of magnitude reasoning. AAAI 1986, p. 100-104. Philadelphia. 1986
21. Raiman, O.: Diagnosis as a trial: The alibi principle. Model Based Diagnosis International Workshop, July 25-27. Paris. 1989
22. Vivet, M., Futtersack, M., Labat, J.M.: Métaconnaissances dans les tuteurs intelligents. Congrès ITS 88. Montréal. 1988
23. White, B.Y.,Frederiksen, J.R.: Qualitative models and intelligent learning environments. Artificial Intelligence and Education. Ablex Publishing Corporation.Norwood. 1987
24. Woods, W.A.: Transition network grammars for natural language analysis. Communications of the ACM vol.13. 1970
25. Yazdani, M.: Intelligent Tutoring Systems : an overview. Artificial Intelligence and Education.Ablex Publishing Corporation.Norwood. 1987

Simultaneous Processing of Different Problem Aspects in Expert Problem Solving

An Analysis in the Domain of Physics on the Basis of Formal Theories of Commonsense Knowledge

Aemilian Hron

Deutsches Institut für Fernstudien an der Universität Tübingen, Konrad-Adenauer Strasse 40-44,
D-7400 Tübingen 1, FRG

Abstract: Expert problem solving in the physical domain is analyzed with respect to simultaneous processing of different problem aspects. On the basis of formal theories of commonsense knowledge in Artificial Intelligence a rational task analysis of a given physical problem leads to a conceptual model of problem solving. This model is a template for an empirical taks analysis by means of analyzing the verbal protocols of four experts. The study reveals that only in one of the experts, the verbal statements could be interpreted to a larger extent in the sense of the information processing searched for. It is assumed that simultaneous processing of different problem aspects leads to a higher redundancy in expert problem solving and thus brings about the chance for mutual control and testing of the single operations.

Keywords: cognitive task analysis, commonsense knowledge, expert problem solving, simultaneous processing, Qualitative Process theory, task network

1. Introduction

Expert problem solving in the domain of physics has been analyzed in several studies, especially in the context of characterizing expert/novice differences [2, 4, 7, 8, 12]. In these studies as well as in other studies of human expertise [3] the superior pattern recognition of experts and their superior capacity to form elaborate chunks in short-term memory has been evidenced. These features are underlying experts' ability to process several aspects of a problem simultaneously. Though this topic should refer mainly to the subsymbolic level of human information processing, it would be of interest to research into it by means of analyzing verbal data in experts' problem solving protocols.

The study is based on a *cognitive task analysis* [11]. Starting point is a *rational task analysis*, that is, a detailed breakdown of the structure of problem solving of a given physical problem. This analysis refers to formal theories of commonsense knowledge in the domain of Artificial Intelligence: Qualitative Process Theory [5, 6] in connection with the theory of

NATO ASI Series, Vol. F 86
Intelligent Learning Environments
and Knowledge Acquisition in Physics
Edited by A. Tiberghien and H. Mandl

reasoning about plans [9, 10]. It leads to a conceptual model of problem solving as a basis for a following *empirical task analysis* of experts' problem solving by means of a protocol analysis. The study is to be regarded as pilot-study which rather serves the formation of hypotheses for future investigations than for hypothesis testing.

2. Theoretical framework

The formal theories of commonsense knowledge, which are chosen as a basis for the rational task analysis, Forbus' [5, 6] Qualitative Process Theory and McDermott's [9, 10] theory of reasoning about plans, are based on detailed assumptions about mental models of physical phenomena and qualitative reasoning.

Qualitative Process (QP-)Theory assumes that for understanding physical objects and situations, the aspect of change is decisive--described in terms of, for example, move, collide, flow, bend, and boil. The other assumption is that such events are intuitively characterized as processes. Commonsense physical reasoning is represented as qualitative reasoning about processes, when they will occur, their effects and when they will stop.

QP-Theory defines a special notion of physical processes and a conception for the qualitative representation of physical parameters. Processes usually start and stop when orderings between object parameters, so called quantities, change, such as unequal temperatures causing a heat flow. These inequalities are represented by means of a quantity space. It is a partial ordering of quantities determined by the domain physics and the analysis being performed.

A process is specified by five parts [6, p. 194]:

(1) *Individuals*: Descriptions of the objects which the process acts on;
(2) *Quantity Conditions*: Inequality statements and status assignments which must be true for the process to be active;
(3) *Preconditions*: Statements other than quantity conditions that must be true for the process to be active;
(4) *Relations*: The relationships between the individuals which hold when the process is active;
(5) *Influences*: Descriptions of the direct effects of the process.

In Fig. 1, a process of acceleration of a system is presented which consists of three masses (M, M1,M2). The system is an element of the problem used in the study (see Fig. 2). The *individuals* point to the fact that the complete system consists of the three masses M1, M2 and

M. The *preconditions* determine the system's free movability, the direction and the presupposed lack of friction. The *quantity conditions* determine the existence of an external

```
process acceleration (TOTAL SYSTEM)

Individuals

M1 a mass
M2 a mass
M  a mass
Total System (M1, M2, M)
dir a direction

Preconditions

Free-direction (TOTAL SYSTEM, dir)
Direction-of (dir, velocity (TOTAL SYSTEM))
Frictionless

Quantity Conditions

A [external force (TOTAL SYSTEM)] > NULL

Relations

Let acceleration, external force and inertial force be quantities
acceleration αQ+ external force
acceleration αQ- masses (M1, M2, M)
acceleration αQ+ inertial force

Influences

I + (velocity (TOTAL SYSTEM), A [acceleration (TOTAL SYSTEM)] )
```

Fig. 1. A physical process definition [5].

force onto the system to be a decisive prerequisite for the coming about of the acceleration process (the letter A shows that the amount of a quantity is decisive, not the derivate). The *relations* characterize the functional dependencies between the involved quantities (a_{Q+} and a_{Q-} indicate that the corresponding quantities are qualitatively proportional in the sense of an increasing or a decreasing monotonic function). The *influences* show that the velocity of the system is directly influenced by the acceleration.

Qualitative thinking about the existence, structure and the effects of processes is described by so-called basic qualitative inferences [6, pp. 197-199]:

(1) *Finding possible processes*: On the basis of a process vocabulary, the possible processes in a physical situation, the so-called process instances, are determined.

(2) *Determining activity*: It is being determined whether or not the preconditions and quantity conditions of these process instances are true. In this way a process instance can have the status active or inactive. The collection of active process instances is called the process structure of the situation. The process structure represents "what's happening" to the individuals in a particular situation.

(3) *Determining changes*: It is being determined which direct and indirect effects are caused by a process. The direct effects of a process onto its quantities are determined in that the respective physical derivation and its sign are determined. The indirect effects are determined on the basis of the so-called qualitative proportionalities, that is, a quantity is specified as a function of another quantity.

(4) *Limit analysis*: The question is how the quantity conditions have to change, in order to start or to end a process. Thus the limit analysis is directed toward determining possible changes in a situation's process structure. It is carried out using the derivate of a quantity and determining how the truth of a quantity condition can change.

(5) *Determining process interaction*: According to the problem one has to infer whether certain processes interact with other processes. This conclusion is central for the understanding of a more complex physical situation which contains several processes. Processes interact by means of common effects. Two processes, which influence the same quantities, or one process, which influences a quantity as to the quantity conditions of another process, have to be considered together, if one wants to find out, whether and how the processes are mutually dependent.

The qualitative inferences are regarded as problem solving operations which come to bear in physical problem solving. According to the theory of reasoning about plans [9, 10], they are regarded as tasks which are included in a task network. A task network structures tasks according to their hierarchy relations and the time-sequence of their processing. Problem solving is considered to be an identification of plans, that is, the breakdown of a problem into subtasks, and their instantiation in the sense of task processing. In the present study, the concept of a task network for structuring of the set of possible problem solving operations comes to bear.

3. Cognitive task analysis

3.1 Physical problem

The study is based on a problem which has already been used by Larkin [7] in a study of problem solving:

Which constant horizontal force F must be applied to the large cart (of mass M) so that the smaller carts (masses M1 and M2) do not move relative to the large cart? Neglect friction.

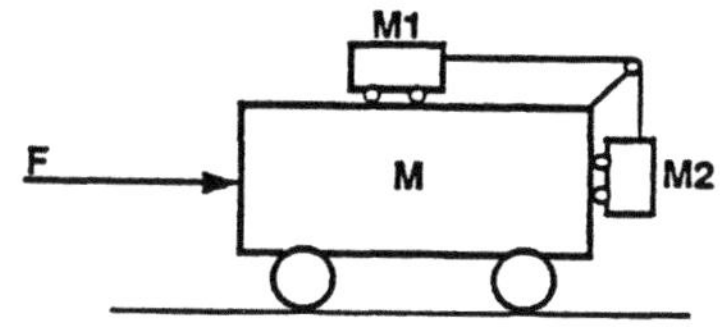

Fig. 2. Illustration of the physical problem [7, p. 81]

4. Rational task analysis

4.1 Physical analysis with respect to basic qualitative inferences

The physical analysis of the problem ensues with respect to the basic qualitative inferences. It can be viewed within two distinct frames of reference. In the first case the problem solver is within an accelerated global system, in the second one he or she is without the system. Only the first case is dealt with, since only that one was considered by the problem solvers in the study.

The starting point of the problem solving is the assumption of two interacting acceleration processes: process acceleration TOTAL SYSTEM (see Fig. 1), process acceleration SUBSYSTEM (i.e. the masses M1 and M2) (Qualitative inferences: *Finding possible processes*). The problem solver considers his or her task to determine the interaction of both processes as to stop the process acceleration SUBSYSTEM.

The problem solver determines the forces acting upon the objects involved in the process acceleration SUBSYSTEM. He or she infers that a gravitational force of M2*g acts upon M2 and that an inertial force of M1*a acts upon M1 (Qualitative inferences: *Determining activity*).

The problem solver now determines the quantity condition of the process acceleration SUBSYSTEM wherein M1 and M2 are motionless in regard to M. This is the case, if gravitational force equals the inertial force. Accordingly, the quantity condition is A [gravitational force (M2)] = A [inertial force (M1)]. In the equilibrium of the subsystem, the gravitational force of M2 is F_G = M2*g and the inertial force of M1 is F_T = M1*a whereby both are considered equivalent: M1*a = M2*g. In accordance with this relation the problem solver determines the acceleration of M1: a = M2*g/M1 (Qualitative inferences: *Limit analysis*).

Now the problem solver specifies the quantity condition of the process acceleration TOTAL SYSTEM, in order to establish the process interaction in the sense of the desired equilibrium condition. On the basis of the supposed resting position of M1, he or she determines the acceleration of the total system (à) as being equal to the acceleration of M1: à = a. The problem solver compensates the assumed inertial force of the total system by means of a force acting from without. He or she determines this force as being the searched for, specific quantity condition of the process acceleration TOTAL SYSTEM:

A [$\text{force}_{(TotSyst)}$] = (M+M1+M2)*a = (M+M1+M2)*M2*g/M1)
(Qualitative inferences: *Determining process interaction*).

Concerning the study, inferences in the sense of *finding possible processes* are excluded. One could start from the fact that these are implicit inferences which can be assessed but indirectly in the prospective problem solvers' verbal protocols. Thus one can assume the following inferences :

- *Determining activity* of process acceleration SUBSYSTEM:
  a) determining of gravitational force of M2: $F_{(G)}$ = M2*g;
  b) determining of inertial force of M1: $F_{(T)}$ = M1*a.
- *Limit analysis* of process acceleration SUBSYSTEM:
  c) determining of equilibrium between gravitational force and inertial force:
  M2*g = M1*a;
  d) determining of acceleration of M1: a = M2*g/M1.
- *Determining activity* of process acceleration TOTAL SYSTEM:
  e) determining of the mass of the total system: $M_{(tot)}$ = M+M1+M2;
  f) determining of inertial force of the total system: $F_{T(TotSyst)}$ = à*(M+M1+M2).
- *Determining process interaction*:
  g) determining of acceleration of the total system: à = a
  h) determining the external force on the total system:
  $F_{(TotSyst)}$ = (M+M1+M2)*a = (M+M1+M2)*M2*g/M1.

4.2 Task network

The qualitative inferences are structured as a task network [cf. 9, 10]. It is presented in Fig. 3 in view of hierarchical relations. According to the above analysis, the problem solution is subdivided into four tasks which split into two subtasks each. The subtask are designated from a to h.

PHYSICS-PROBLEM							
Determining Process Activity SUBSYSTEM		Limit Analysis SUBSYSTEM		Determining Process Activity TOTAL SYSTEM		Determining Process Interaction SUBSYSTEM/ TOTAL SYSTEM	
Deter-mining Gravitational Force	Deter-mining Inertial Force	Deter-mining Equilibrium Gravitational Force/ Inertial Force	Deter-mining Acceler-ation M1	Deter-mining Mass TOTAL SYSTEM	Deter-mining Inertial Force TOTAL SYSTEM	Deter-mining Acceler-ation TOTAL SYSTEM	Deter-mining External Force on TOTAL SYSTEM
a	b	c	d	e	f	g	h

Fig. 3. Task network in view of the hierarchical relations

In addition to the hierarchical relations, the precondition relations are essential. They relate, for example, to the fact that the task *determining process interaction* can not be done unless the *limit analysis* of process acceleration SUBSYSTEM has been performed. A systematic consideration basing on the hierarchical relations, the precondition relations between the superordinated tasks, as well as the specific preconditions of each respective pairs of subtasks leads to the task network presented in Fig. 4. This task network is a directed graph of possible sequences of performed tasks on the time-line. Each node of the graph represents a stage in course of working on the tasks. The starting point is x_0. Now, either task b or task e can be processed. If one works on task a, then stage x_{11} reveals that afterwards either task b or task e can be processed. If one works on task e, then stage x_{22} shows that the subsequent possible tasks are either b or f, and so forth. In this way thirty different pathways result on which the graph can be processed. Thus there are thirty different possibilities of task sequencing in problem solving.

The task network in the form presented in Fig. 3 and Fig. 4 is a conceptual model for the empirical analysis of experts' problem solving.

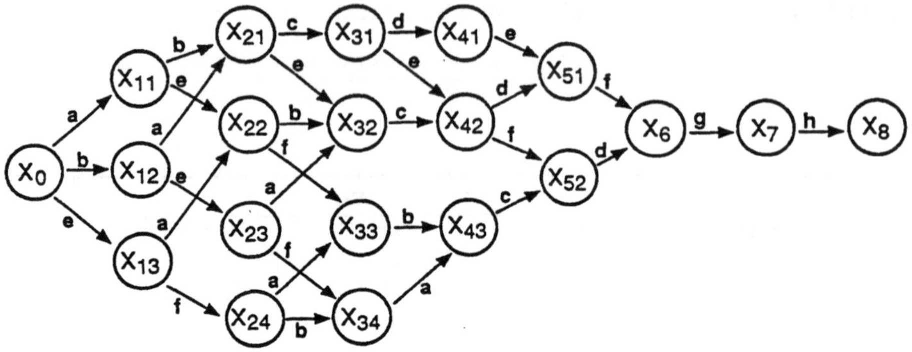

Fig. 4. Directed graph of possible sequences of tasks a through h

5. Empirical task analysis

5.1 Method

The empirical study is based on the analysis of verbal protocols of experts' problem solving grounded on the developed conceptual model. In a prior study, the verbal protocols of two experts were analyzed. Thereby it was revealed to be necessary to supplement the recognized qualitative inferences. It was evidenced that the experts in a general way reflected on the relationship of the total system with the subsystem as well as on the subsystems' acceleration process before having performed the described specific inferences (additional categories "Constructing Relations TOTAL SYSTEM" (y), "Constructing Relations SUBSYSTEM" (z)). Furthermore, it was shown that the experts repeatedly recalled the conditions of the task (additional category "memorizing"). Finally, a category for other content ("rest") was formed. All of the categories are shown in the head of Table 1.

In the main study, the verbal protocols of 4 problem solvers (3 PhD physicians, 1 student about to take physics exam in the 12th semester) were examined. Beforehand, the problem solvers were introduced into the method of thinking aloud. The verbal statements were recorded and transcribed per line. The lines of the transcript were parsed to form statements according to the developed categories [cf. 1, for the method]. Examples are shown in Table 1. A control evaluation by a trained evaluator resulted in an interrater reliability of .68. However, the verbal protocols were already parsed in statements which were only to be categorized by the second rater.

5.2 Results

The problem was solved by all experts. However, two experts produced a wrong partial solution, because the mass of the total system was erroneously regarded to be M and not as a collection of M, M1 and M2.

The solution path of every one of the four problem solvers (Pb1, Pb2, Pb3 and Pb4) can be viewed in Table 1. For every problem solver, each of the statements under the respective categories (see top row of table) are connected by lines. Four solution graphs result which are presented below one another. Dotted lines in the solution graph mean that statements without direct contentwise relations to the qualitative inferences have been performed (categories "rest", "memorizing"). The hatched areas wherein the solution graphs of Pb1, Pb2, and Pb3 lie in part, represent incorrect problem representations.

Table 1 reveals that, on average, there were 18 to 19 statements per expert. In only one of the tested experts (Pb4), the verbal statements can be interpreted to a larger extent in the sense of the assumed simultaneous problem solving behavior. It refers to the second, third, fourth and eleventh statement of Pb4. This expert attracts attention due to his smooth and efficient problem solving and his correct problem representation from the onset of the problem solving process.

For the purpose of illustration in Table 2, the first part of the verbal protocol of Pb4 is represented. The first statement is an inference concerning the total system and points out his immediate correct problem representation. The next three inferences are of interest with respect to the experts' reasoning which is searched for: They contain several problem aspects (tasks) in an intertwined way. In his second statement, Pb4 identifies the respective forces in the subsystem (categories a, b) and the need for their equilibrium (category c). The third statement also shows a simultaneous consideration of some problem aspects: Pb4 considers the acceleration of the subsystem, resulting from the equilibrium (category c), to be the acceleration of the total system (category g). This idea is fundamental for solving the problem! It is taken up in the forth statement (category g) in connection with determining the acceleration of the subsystem (category d).

Table 1: Solution paths

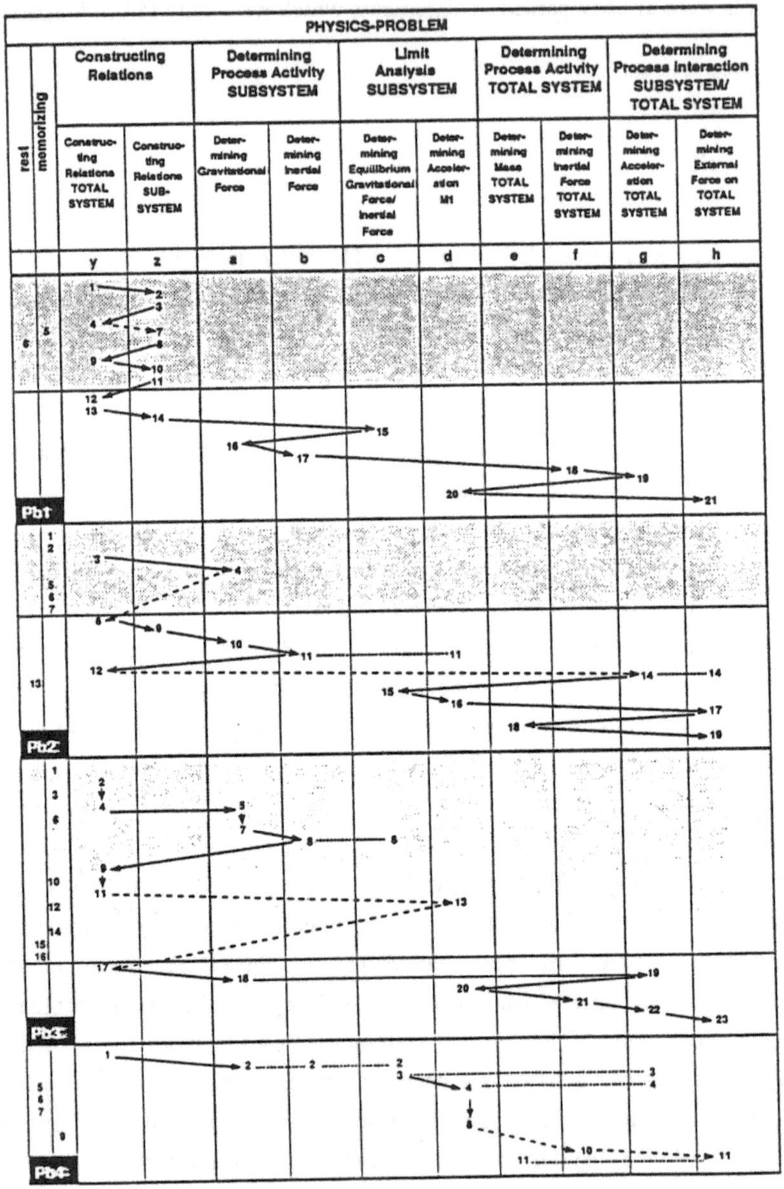

Table 2: Partial transcript of a protocol parsed in statements

No.	category	content of statement
1	y	Mmh, das ist jetzt gar nicht so einfach. Sobald sich nämlich der untere Wagen bewegt, treten ja an den beiden kleinen Massen Trägheitskräfte auf, [Hm, that's not that single. For, as soon as the lower cart moves, inertial forces are being exerted at the small ones,]
2	a / b / c	so daß ich jetzt einmal überlegen muß, wie groß hier die Kraft an der Masse M1 werden muß, damit die Trägheitskraft an der Masse M1 gleich der Gewichtskraft M2 wird. [thus I now have to ponder on how great the force exerted at mass M1 has to be so that the inertial force exerted at mass M1 equals the gravitational force exerted at M2.]
3	c / g	Das wäre also die Kraft an der Masse M1 gleich M1 mal die Beschleunigung, die ich dem System (Gesamtsystem; A.H.) geben muß, mal der Beschleunigung a, und das muß gleich sein der Masse M2 mal der Erdbeschleunigung. [That would thus be the force exerted at mass M1, that is, M1 times the acceleration a, the acceleration I have to confer to the system (total system; A.H.), times the acceleration a, and that should be the same as mass M2 times the Earth gravity.]
4	d / g	So, daraus folgt also, aus diesem Ansatz, daß die Beschleunigung aus dem Verhältnis der beiden Massen (M1, M2; A.H.) mal der Erdbeschleunigung sich zusammensetzt ... [Thus, out of this approach the conclusion results that the acceleration is composed of the relation of both masses (M1, M2; A.H.) times the Earth gravity]

Similar merging of different problem aspects prevail in the eighth statement of Pb3 as well as in the eleventh and fourteenth statement of Pb2 (see Table 1). In all of the other subjects' statements, only one problem aspect (task) at a time is considered. It should be noted that the observed interlacing of different problem aspects in some cases involves a higher redundancy of the problem solving operations (see Table 1).

A comparative analysis was undertaken with respect to the sequence of subjects' inferences by means of the graph constructed in the rational task analysis (see Fig 4). It reveals little correspondence concerning the three subjects, Pb1, Pb2 and Pb3. In contrast, Pb4 shows a better correspondence with the logical sequence of the solution operations. It is probably due to the fact that Pb4 - in contrast to the other subjects - does not reflect much beforehand, but starts immediately the solving process in the right way.

6. Discussion

The study points out that only in one of the tested experts the verbal statements could be interpreted to a larger extent in the sense of the assumed simultaneous processing of different aspects of a problem. Admittedly, the protocol analysis used in this study is not optimal for analyzing this phenomenon. Simultaneous processing in problem solving refers, at the first instance, to the subsymbolic level of pattern recognition and chunking which should be studied in connection with methods such as measures of reaction time. Moreover, the result of the study has been found by means of interpretation. Nevertheless, the protocol analysis furnishes some hints about the procedure of experts' problem solving: It could be supposed that simultaneous processing of different problem aspects leads to a higher redundancy of problem solving operations. Thus the chance for mutual control and testing of the single operations is brought about leading to a smooth and efficient experts' problem solving. This topic should be investigated in future studies.

References

1. Chi, M.T.H., Bassok, M., Lewis, M.W., Reimann, P., & Glaser, R.: Self-Explanations: How Students Study and Use Examples in Learning to Solve Problems. Cognitive Science, 13, 145-182, 1989
2. Chi, M.T.H., Feltovich, P., & Glaser, R.: Categorization and Representation of Physics Problems by Experts and Novices. Cognitive Science, 5, 121-152, 1981
3. M.T.H. Chi, Glaser, R., & Farr, M.J. (Eds.): The Nature of Expertise. Hillsdale, N.J.: Lawrence Erlbaum, 1988
4. Chi, M.T.H., Glaser, R., & Rees, E.: Expertise in Problem Solving. In: Advances in the Psychology of Human Intelligence (R.J. Sternberg, ed.), Vol. 1, pp. 7-75. Hillsdale, N.J.: Lawrence Erlbaum, 1982
5. Forbus, K.D.: Qualitative Process Theory. In: Artificial Intelligence (K.D. Forbus, ed.), 24, 85-168, 1984
6. Forbus, K.D.: The Role of Qualitative Dynamics in Naive Physics. In: Formal Theories of the Commonsense World (J.R. Hobbs & R.C. Moore, eds.), pp. 185-226. Norwood, N.J.: Ablex Publishing Corporation, 1985
7. Larkin, J.H.: The Role of Problem Representation in Physics. In: Mental Models (D. Gentner & A.L. Stevens, eds.), pp. 75-98. Hillsdale, N.J.: Lawrence Erlbaum 1983.
8. Larkin, J.H., McDermott, J., Simon, D.P., & Simon, H.A.: Models of Competence in Solving Physics Problems. Cognitive Science, 4, 317-345, 1980
9. McDermott, D.: A Temporal Logic for Reasoning About Processes and Plans. Cognitive Science, 6, 101-155, 1982
10. McDermott, D.: Reasoning about Plans. In: Formal Theories of the Commonsense World (J.R. Hobbs & R.C. Moore, eds.), pp. 269-317. Norwood, N.J.: Ablex Publishing Corporation, 1985
11. Resnick, L.B. & Ford, W.W.: The Psychology of Mathematics for Instruction. Hillsdale, N.J.: Lawrence Erlbaum 1981
12. Simon, D.P. & Simon, H.A.: Individual Differences in Solving Physics Problems. In: Children's Thinking: What Develops? (R.S. Siegler, ed.), pp. 325-348. Hillsdale, N.J.: Lawrence Erlbaum 1978

Modelis : an Artificial Intelligence System Which Models Thermodynamics Textbook problems

G.Tisseau

LAFORIA, Université Paris 6, 4 rue Jussieu, F-75252 Paris Cedex 05, France

Abstract: When a problem in physics is stated in concrete terms, in an informal and incomplete way, students have difficulties to solve it, because the theory that they learn in the classroom does not match with the terms of the problem. Modelis is an Artificial Intelligence system which deals with this difficulty in the domain of Thermodynamics. It starts from texts in natural language, extracted from textbooks (in French), and builds successive representations of the problem, called linguistic, concrete and theoretical. With the theoretical representation, formal knowledge is applied to produce a plan of the solution, by propagating information in a constraint network. This paper describes the general architecture of the system, discusses problems which appear in the modelling phase (instability, incompleteness, identification), presents some methods of modelling (simplification, passage from local to global, transfer of information, reformulation) and shows how the plan of solution is obtained.

Keywords: artificial intelligence, modelling, problem solving, knowledge representation, change of representation

1. Presentation of Modelis

Modelis is an Artificial Intelligence system which solves textbook problems in Thermodynamics. Actually, its main purpose is not really to find the solution, but to build a theoretical model of the problem, which is initially stated in concrete terms, in an informal and incomplete way. The students have difficulties with this sort of problem because the theory that they learn in the classroom does not match with the terms of the problems. The theory is about conceptual entities (system, interaction, pressure, transformation), whereas the problems are about physical objects and events (a cylinder, a thrust on a piston).

The elaboration of Modelis followed a rather empirical process, with no particular psychological theory in mind. The general idea was to find out the knowledge and processes necessary to make it work on a computer. Once this knowledge is made explicit, it can be a

NATO ASI Series, Vol. F 86
Intelligent Learning Environments
and Knowledge Acquisition in Physics
Edited by A. Tiberghien and H. Mandl

basis for a further research : it can be used to build an intelligent tutoring system, it can be tested on students and experts to see if it has a psychological relevance, it can be generalized to other domains.

A similar work was done in mechanics by Bundy with the system Mecho [1], but the distinction between the representations was not so clear, and the techniques used were different. The presence of different representations was observed on experts and novices by Chi et al [2], and the importance of a good representation is discussed by Larkin [6].

The informal description of the problem could be said to correspond to a "naive theory" of physics, that is to say a set of conceptions and beliefs about the physical world. These conceptions are not necessarily false, but they are not of a scientific nature and so are not at the appropriate level to solve a problem. They only serve to describe the reality in terms of observed objects and events. To be able to apply the theory in order to solve a problem, it is necessary to have a modelled description of this problem. This description must be of a scientific nature : complete, unambiguous, and using only the concepts of the theory.

The distance is often important between the informal description and the modelled one, and the task to build the modelled one from the informal one is not a direct translation (this is pointed out by Reif [8]). So it is necessary to introduce an intermediate description, which can evolve gradually from the initial text to the modelled one. Actually, several intermediate reresentations can be distinguished in Modelis : they can be called linguistic, concrete, and theoretical. They are presented in the next section.

2. The different representations

2.1 Linguistic representation

When a student reads a text, he must first understand it on a linguistic level : he must recognize the words, the groups of words, their relationships. To do that, he needs linguistic knowledge, which can be general (syntactic structures) or specific (particular words and expressions used in the domain).

For example, one of the sentence is : "A mass of gas enclosed in a cylinder with a diameter of 200 mm occupies a height of 500 mm".

To build a linguistic representation, it is necessary to recognize the principal words and their meaning (mass, gas, enclosed , ...), and to decompose the sentence into simple relationships :

a mass ... occupies a height ...

a height of 500 mm
a mass of gas
a mass ... enclosed in a cylinder
a cylinder with a diameter ...
a diameter of 200 mm

This decomposition is done in Modelis by a linguistic analyzer which will be presented later. It needs not only syntactic knowledge, but also domain knowledge. For example, it is necessary to recognize broad categories of entities (matter, device, quantity, event, ...) and to know how they can be related to each other (a quantity can be applied to a matter, a device can be connected to another device, etc...).

2.2 Concrete representation

The linguistic representation is not enough to really understand the text. It is just a literal translation. It is now necessary to form the equivalent of a mental representation of the experiment that is described by the text. For example :

"The dimensions of a gasometer bell are : diameter 6 m, height 6 m. Its mass is 6200 kg. What will be the emerging height when the gasometer contains 100 m3 of gas ?"

The student must know what a gasometer is : it is composed of a tank containing water, and a bell enclosing gas. He must know that there are two levels of water (internal level and external level), and what "emerging height" means (distance between the top of the bell and the external level). Knowing this, he is able to draw a diagram of the apparatus, or to imagine the scene, or even to recreate the experiment.

This can be called a concrete representation, because it is centered on the physical objects, the topological structure, and the observed events. It corresponds to the immediate perceptions. It is useful because it has some knowledge attached to it : the naive or concrete knowledge which helps to solve ambiguities and to complete what was left implicit in the text. Modelis constructs a concrete representation and uses such knowledge to transform it.

2.3 Theoretical representation

The concrete representation is not enough to solve the problem. This can be done only by using the theory. But the concepts of the theory are not necessarily related to concrete objects, but also to physical principles. In the last example, the important concepts to introduce are the thermodynamical systems corresponding to the gas and the atmosphere. The other objects (the water and the bell) can be considered as intermediaries, materializing interactions between the two main systems. With each of these systems or interactions are associated physical principles that can be used to solve the problem : the atmosphere and the gas verify the laws of thermodynamical equilibrium, the interaction through the water can be studied with the laws of statics of fluids, and the interaction through the bell by the laws of mechanical equilibrium.

This representation is centered on systems, interactions, transformations, and theoretical principles. These entities are rarely mentioned in the text of problems, but it is necessary to evoke them to solve the problem. Students find it difficult to evoke the proper concepts.

Modelis builds a theoretical representation and uses it to solve the problem, using methods which will be discussed later .

3. Architecture of Modelis

The Modelis system uses these different representations (linguistic, concrete, theoretical). It works in two main phases : first a linguistic analysis and then a modelling and solving phase.

3.1 Linguistic analysis

The linguistic analysis starts from the text of an exercise, given in natural language (in French), as it appears in a textbook. The text is first preprocessed (elementary grouping of words, rewriting of some words or expressions), then it is processed by an analyzer to yield the linguistic representation.

The *analyzer* has a general syntactic knowledge base (dictionaries, description of syntactic structures). It works in a combinatorial way, exploring the different structures in a descending manner to finally match the text. But it has no specific knowledge on a particular domain, and so this exploration would lead to many possible analyses. Most natural language systems would then examine each of these possibilities, using semantic knowledge in order to reject the inconsistent ones and finally find the right interpretation. It is not the case in Modelis, where the semantic analysis is made concurrently with the syntactic one, by means of a module named "interpreter" which uses "interpretation rules".

The task of the *interpreter* is primarily to build the concrete representation of the problem from the linguistic representation. At each step of the syntactic analysis, pertinent rules are examined by the interpreter, depending on which syntactic event happened. These rules are production rules with conditions and actions. The conditions can be syntactic or semantic ones : they can examine both the linguistic representation and the concrete representation, which are build concurrently. The actions can build the semantic interpretation or detect inconsistencies, which can force the syntactic analyzer to backtrack to another possibility.

So there is a constant interaction between the analyzer and the interpreter : the analyzer builds the linguistic representation and calls the interpreter at each step, and the interpreter builds the semantic interpretation and redirects the analyzer.

At this stage, few or no theoretical thermodynamical concepts are introduced. The representation is only a description of objects and events.

We will not focus in this paper on the linguistic analysis. Natural language processing by computers has been extensively studied, and there is still active research in the field. The linguistic analyzer of Modelis uses some classical techniques (the syntactic analyzer) with some extensions (the interaction of the analyzer and the interpreter by means of interpretation rules). It is not intended to be general, but seems well suited for the kind of texts studied. A similar work, with different methods, is that of Mellish [7], for texts in the domain of mechanics.

3.2 Modelling and solving phase

The concrete representation is passed to a second subsystem, and will evolve through an intermediate representation to finally yield a theoretical representation, and a plan of solution. It is not possible to describe here in detail the computer techniques used to implement Modelis : a complete description can be found in [9]. The core of the system is an inference engine specially designed for Modelis, which uses production rules with variables.

Some production rules form a module called "modeller", which modifies and completes the intermediate representation, makes hypotheses, and generates the theoretical representation. Another module called "solver" uses this theoretical representation, and formal knowledge in Thermodynamics, to construct a plan of the solution : which quantities to compute, in which order, and with which equations. To obtain this plan, the equations are represented in a qualitative way as constraints forming a network, and the information is propagated in this network.

To give an idea of what sort of tasks are involved, let us see the different representations associated with a piece of text : "the gas is compressed to 2 bars".
After the linguistic analysis, an internal object M is created, corresponding to the concept of measure :
M is a measure [object: the gas, value: 2 bars, operator: final-value, simultaneous: compression]
It represents the following facts :

the gas has been measured,

the value found is 2 bars,

this value is the final value in a certain period of time,

this period is the same as the one of the compression.

But this is not a complete description, and it is not entirely stated in the terms of the theory. The modeller must then create new objects and complete the object M.

A thermodynamical system must be introduced, associated with the gas (after identifying this gas, presumably mentioned earlier in the text) :

create a system S [matter : the gas].

A period of time must be introduced, with two instants :

create a period P [initial-instant: I, final-instant: F].

This period P is associated to the compression :

period(compression) = P

Now the measure M can be modified and completed :

M is a measure [system: S, period: P, instant: F, quantity: pressure].

This form is quite different from the initial one. It can be used to generate a piece of the theoretical representation :

p(S F)

This is a predicate meaning that the pressure of the system S at the instant F is known. The information is only qualitative (the value is not mentioned). The whole theoretical representation is build by such processes.

4. Some problems

4.1 Problem : incompleteness

The information given in a textbook problem is often insufficient for several purposes : to represent the device which is described, and also to solve the problem.

Insufficient information to represent the device : for example, let us consider the sentence : "A container contains oxygen". This description is incomplete if we want to represent the device accurately : what is the shape of the container ? Is it closed ? Where is it placed (in the atmosphere) ? Are its sides isolating ? Is the oxygen a gas or a liquid ? All these facts could be important, depending on the question.

Insufficient information to solve the problem : in the following example, the information is not sufficient to solve the problem :
"A chimney with a height of 60 m receives 4000 m3/h of smoke at 250°C. Its temperature decreases of 0.6 °C per meter. What is the volumetric rate at the output ?"
If we want to solve the problem, we must answer the following questions : what is the law of behaviour of the gas (perfect gas ?) ? What are the pressure conditions (constant pressure ? with which value ?) ? This information are not given in the text, so we must make hypotheses.

In Modelis, there are rules that allow to complete this information. For example, if the pressure conditions in a factory chimney are not known, assume that the pressure is constant.

4.2 Problem : instability

Many rules in the system are default rules, which say that some properties can be assumed in case of missing information. For example :
"If there is no transformation mentioned, then there is only one instant to consider."

The presence of a transformation is tested not only in the explicit contents of the text, but also in the representation, which is gradually build by inferences. It may be that at a certain stage, there is no transformation in the representation, so the rule is applicable, and the system should conclude that there is only one instant to consider. But at a later stage, it may be that a transformation is present, and so the rule would not be applicable. The representation would be inconsistent.

This example shows that some states of the knowledge base are not meaningful enough to apply certain rules. To avoid this difficulty, two methods are used :
1) Rules can be grouped in modules called "expertises", which can be examined sequentially, in an order determined by the author of the rules. It is then possible to examine default rules only after examining the expertises that ensure that the knowledge base is in a meaningful state.
2) If the first solution is not practicable, it is possible to incorporate in the default rules themselves supplementary conditions preventing their wrong application.

4.3 Problem : identification

During the construction of the modelled representation, some objects are evoked, created, and introduced in the representation. A problem can then arise : two of these objects, evoked for

different reasons, can in fact represent the same entity, and it is then necessary to identify these objects. The following example shows this difficulty :
"A cylinder with a diameter of 200 mm contains oxygen at 15°C, occupying a height of 500 mm. The piston is rapidly driven. The pressure varies from 1 bar to 2 bars. What is the variation of the internal energy ?"
Several situations described here evoke the presence of concepts to be introduced in the representation :
"A cylinder contains oxygen at 15°C" evokes a first object S1, an equilibrium state of the oxygen.
"The pressure varies from 1 bar to 2 bars" evokes two other objects S2 and S3. S2 is an equilibrium state of a system (corresponding to 1 bar), and S3 is also an equilibrium state of a system (corresponding to 2 bars). But how many distinct concepts do the three objects S1, S2 and S3 represent ? In fact, only two : the initial state of the oxygen (S1 and S2), and the final state (S3). To understand the text, it is necessary to detect that S1 is the same as S2.

In Modelis, two methods are used to handle this problem.
1) The first method is not to create the objects as soon as they are evoked, but to refer to them indirectly. For example, it is possible here to represent the fact that in the initial state, the temperature is 15 °C, without having to create an object representing the initial state. It is also possible to detect that the two measures (15 °C and 1 bar) are simultaneous. When an equilibrium state is needed in the representation, this information will permit to create only one state instead of two (S1 and S2).
2) The second method is to state rules that detect such identifications and that replace two objects representing the same entity by only one object, combining the properties of the two. This is similar to what is called "resolution of references" in natural language processing.

5. Methods of modelling

In this section, we present some of the methods used by Modelis during the modelling phase.

5.1 Simplification

To transform a concrete representation into a modelled one, it is often necessary to make some simplifications. The usual way to simplify is to forget something.

Forget properties : when the text says "a compressed gas bottle", that is to say a bottle which usually contains compressed gas, it is possible to forget the details and to read only "a bottle". When the text says "a city gas", that is to say a gas which is usually distributed in cities, it is possible to read only "a gas". This is achieved in Modelis by means of rewriting rules, which are applied before any analysis of the text.

Forget objects : the text often describes an experimental device, but the particular apparatus composing this device are here only to introduce constraints on a fluid which is processed by the device. So it is not always necessary to represent the apparatus explicitly. For example, in the following text : " A closed tank contains some air at 35 °C and under a pressure of 7 bars. What will be the pressure when the temperature falls to 10 °C ? ", it is possible to forget the tank in the abstract model. The important entity is the air.

Neglect phenomena : in a physical experiment, several phenomena can in principle be observed, but their effects are often negligible. An expert knows the existence of these phenomena and knows under which conditions he can neglect them. A novice may be confused when he knows the existence of a phenomenon and doesn't know that he can neglect it. For example, in ordinary situations, the following phenomena can be neglected : the expansion of a container, the buoyancy in a gasometer, the heat transfer in a brief phenomenon, the pressure variation in a factory chimney.

Modelis applies default rules to neglect some phenomena when they are not explicitly mentioned.

5.2 Passage from local to global

The concrete representation obtained after the linguistic analysis is often based on local information. When it is possible, it is better to merge multiple local information into a global one, in order to evoke higher order knowledge. For example, we can learn that there is an input flow of gas into a component, and also an output flow of the same gas out of the same component at the same time. These two flows can then be considered as just two local views of a global flow through the component. This global view can bring more information than just the union of the two local ones, because there is supplementary knowledge attached to it (for example, the law of conservation of matter).

5.3 Transfer of information

It is often a good method to express information at a high level of description, in order to be able to manage the complexity of a problem. A case where it is possible to do that is the case of a system composed of several parts. Information about a part can be converted in information about the whole, thus enabling the application of knowledge associated to the whole system. For example, if a text mentions the diameter of the bell of a gasometer, this information is first represented as a measure :
measure[object : the bell, quantity : diameter]
But if we want to apply general knowledge about gasometers, it is better to express it as a measure applied to the gasometer :
measure[object : the gasometer, quantity : diameter-of-bell].
The information is thus transferred from the part to the whole. This is done by rules in Modelis.

In Modelis, there is a knowledge module about gasometers able to handle this sort of information : if the diameter is known and if the volume of the gas is known, then the height of the gas is known, etc... This module expresses knowledge directly related to a system seen as a whole (the gasometer), without describing its parts (the bell, the tank, the water).

5.4 Reformulation

In the process of modelling, the concrete description must be converted into a theoretical one. In the concrete description, there can be information about a technological structure. For example, consider a cylinder containing a gas and placed in the atmosphere. A precision about the structure can be that the cylinder is closed by a piston. This is represented by a computer structure : closing [container : the cylinder, lid : the piston], which says that there is a situation of closing (a container is closed by a lid), where the role of the container is played by the cylinder and the role of the lid is played by the piston.

In the theoretical representation, the topological structure is no more relevant. The important entities are the systems and the interactions. The central system is the one associated with the gas. Its general behaviour can be described by its type : gas-in-cylinder, corresponding to a module of knowledge. The structural information about the piston, playing the role of lid, must be converted into a theoretical information about the system : the system is closed. Although the same word "closed" is used in the two descriptions (concrete and theoretical), it doesn't mean the same thing. In the concrete sense, it means that there is a particular topological arrangement between concrete objects, whereas in the theoretical sense, it means that there is no transfer of matter between the system and its environment. Theoretical knowledge about closed systems can now be applied.

6. Resolution

6.1 Theoretical representation

The theoretical representation is based on the fundamental concepts of physics : system, state, transformation, interaction, transfer. The properties and parameters associated with these concepts are expressed by predicates, as in formal logic. For example, the property of being a perfect gas is denoted by *pg*, and to state that a system s is a perfect gas, we say : *pg(s)*. To say that the molar mass of s is known, we say : *M(s)*.

At different instants, a system is in different states. Each state can then be designated by a couple (system, instant). A state can have properties and parameters attached to it. For example, to say that the system s is a perfect gas at equilibrium at the instant i, we say : *equiPG(s i)*. To say that the pressure of the system s at the instant i is known, we say : *p(s i)*.

Between two instants, the system undergoes a transformation, which can be designated by a system and two instants. A transformation can have properties and parameters too. For example, to say that the system s undergoes an isothermal transformation between the instant i and the instant f, we say : *isothermal(s i f)*. The amount of work received by the system is denoted by *W(s i f)*, the variation of pressure by *D(p s i f)*.

There can be several systems interacting together. At each instant, there can be relationships between the two systems expliciting this interaction. For example, to say that the two systems s and s' have the same pressure at the instant i, we say : *interRel(p s s' i)*.

There can be transfers of energy between two systems. To say that the work received by system s from system s' between the instants i and f is known, we say : *Wexchanged(s s' i f)*. This formal language enables the writing of rules, and their grouping into modules. A module is a group of rules associated with a predicate. For example there is a module associated with the predicate *equiPG(s i)*, grouping all rules that apply when a system s is a perfect gas at equilibrium at the instant i.

6.2 Constraints

Once the theoretical representation is obtained, the task of the system is to solve the problem. Modelis has no mathematical computation capabilities, so the solution is not presented as a numerical value or a formula, but as a plan showing the quantities to compute, and the equations used to compute them. To obtain such a plan, Modelis uses a qualitative representation of equations. For example, the equation m = M* n, applied to a system s at an instant i, is represented by *rel(m(s i) n(s i) M(s))* , which only says that there is a relation between m(s i), n(s i), and M(s), without precising the exact form of this relation. In order to build a plan of solution, the only thing to know is that any of these three quantities can be computed if the two others are known.

The different equations corresponding to the problem are constructed dynamically, from the theoretical representation. For example, if s is a perfect gas at equilibrium at the instant i (which is represented by *equiPG(s i)*), then the equation *rel(p(s i) V(s i) n(s i) T(s i))* is constructed, corresponding to the law of perfect gases pV = nRT. This construction results from the knowledge module associated with the predicate *equiGP*. The equations act as constraints between the different quantities, connecting them to form a network.

Consider for example the network resulting from the following properties : s is a system at equilibrium at the instants i and f, and s stays closed between i and f.

This is represented by : *system(s) equi(s i) equi(s f) closed(s i f)* . The knowledge associated with the predicates *system*, *closed*, and *equi* is used to construct two constraints C1 and C2 (among others) : C1 is the relation m = nM between the mass m, the molar mass M and the number of moles n : *rel(n(s i) m(s i) M(s)),* and C2 is the conservation of the number of moles *constant(n s i f)*. These constraints create a network connecting the four quantities n(s i), M(s), n(s i) and n(s f), initialized with an unknown value. Depending on the information given in the problem, some of these quantities happen to be known, and the others are then found by propagation in the network. Constraint propagation is a powerful mechanism, used for example in qualitative physics [4,5]. For example, let us assume that M(s) and n(s f) are known. The constraint C2 ensures us that n(s i) can be computed, and then the constraint C1 ensures us that m(s i) can be computed.

The equations could be represented by rules, but then each equation would be associated with several rules : m = n M would correspond to three rules :

if m and n are known then M is known;

if M and n are known then m is known;

if m and M are known then n is known.

Here, this equation is represented by just one constraint, and the general mechanism associated with every constraint allows the system to act as if it applied the three rules.

6.3 Results

At the present time, Modelis solves twenty problems extracted from textbooks, at the first level of university (18 or 19 year old students), in the domain of perfect gases. The length of a text can vary from 12 words to 100 words.
Let us see an example that shows the result of a session with Modelis.
The initial text is :
"A closed tank contains some air at 35°C and under a pressure of 7 bars. What will be the pressure when the temperature falls to 10 °C ?"
After the linguistic analysis, in the modelling phase, Modelis creates three theoretical objects : a system s and two instants i1, i2, corresponding respectively to the air, the initial state, and the final state. The modelled text finally obtained is :

perfectGas(s)
composition(s)
T(s i1)
p(s i1)
T(s i2)
transPGclosed(s i1 i2)
constantV(s i1 i2)
asked(p(s i2))

It says that s is a perfect gas whose composition is known, which undergoes a transformation between the instants i1 and i2 with no transfer of matter and with a constant volume. The initial pressure and temperature are known, and the final temperature too. The final pressure is asked.
To obtain this theoretical representation, the text had to be completed and some hypotheses made : "the pressure" is the pressure of the air in the tank, the air can be considered as a perfect gas, the capacity of the tank can be considered as constant.

With this new formulation, a network of constraint is constructed, and a plan of the solution is finally obtained, which is a list of quantities to compute, or properties to establish, with some explanations (the comments in parentheses are not generated by the system) :

constantV (s i1 i2)	given	(the volume is constant)
inv2(p T s i1 i2)	p/T is constant	(invariance relation)
T(s i1)	given	(initial temperature)
p(s i1)	given	(initial pressure)
law2(p T s i1 i2)	p/T = p1/T1	(the law of dependance is now known)
T(s i2)	given	(final temperature)
p(s i2)	p2/T2 = p1/T1	(use the law of dependance to compute p2)

7. Conclusion

The implementation of a system like Modelis which solves problems stated in concrete terms has several implications for education and intelligent tutoring systems.

First, it shows that in the process of understanding and solving a problem, it is advantageous to have different representations. For the kind of problems considered here, Modelis uses three levels : linguistic, concrete and theoretical. Each type of representation has some specific knowledge attached to it, which enables to complete and modify the representation, and to generate the next representation. Students often fail to solve a problem because they have not the necessary knowledge to generate a representation from another, and so they cannot connect the initial statement of the problem to the theory that would be needed to solve it.

Then, it shows that the knowledge necessary to build a theoretical model from a concrete description is rather pragmatic and uncertain. For example, it is often necessary to make plausible hypotheses, to forget objects and properties. This knowledge is not of the same nature as the formal scientific knowledge used in the theory : it cannot generate rigorous proofs based on axioms and theorems. It is not absolute, but only applicable in usual situations. It is often ignored in the classroom, because it is more difficult to be expressed precisely and formally than a quantitative theory.

One of the benefits of implementing an Artificial Intelligence system is to make explicit a body of knowledge able to solve a given type of problem. Maybe this knowledge is not "the right one" in terms of psychological or didactical criteria, but it is a basis for further development. One possible and intended development is to extract from the current rules general modelling methods, applicable in different domains, not only in Thermodynamics or even in Physics. For example, a modelling phase is necessary in many combinatorial and probability problems. A modelling capacity is essential for an Artificial Intelligence system if it has to solve problems from the real world.

Another possible development would be to build an intelligent tutoring system intended to help students to carry out modelling tasks. The student would have to build a model from a concrete description, and the system could guide and control the elaboration of this model, by giving advice, explanations and refutations. This would show the students that the theoretical knowledge is not the only one to be used, and that modelling is an important phase in problem solving.

References

1. Bundy A. , Byrd L. , Luger G. , Mellish C. , Palmer M. Solving mechanics problems using meta-level inference. Procs of IJCAI 79, pages 1017-1027, 1979
2. Chi M.T.H. , Feltovitch P.J. , Glaser R. Categorization and representation of Physics problem by experts and novices. Cognitive Science 5, 121-152, 1981
3. De Kleer J. , Brown J.S. A qualitative physics based on confluences. Artificial Intelligence 24 (1984) pages 7-83
4. Forbus K.D. Qualitative process theory. Artificial Intelligence 24 (1984) pages 85-168
5. Gentner D., Stevens A.L. (eds). Mental models. Lawrence Erlbaum Associates 1983
6. Larkin J.H. The Role of Problem Representation in Physics. in [5]
7. Mellish C.S. Computer interpretation of natural language descriptions. Ellis Horwood Limited 1985
8. Reif F. Interpretation of scientific or mathematical concepts : cognitive issues and implications. Cognitive Science 11, pp 395-416, 1987
9. Tisseau G. Modélisation à partir d'un énoncé informel : le système Modélis. Application à des exercices de thermodynamique. Thèse d'université, université Paris 6, France.

2. Different Approaches to Student Modelling

Steps Towards the Formalisation of a Psycho-logic of Motion

Joan Bliss[1], Jon Ogborn[2]

[1]King's College London, University of London

[2]Institute of Education, University of London

Abstract: We propose some steps towards a formalisation of a commonsense theory about motion, more particularly its development. This account attempts to show first how a number of primitive actions, basic schemes and rules evolve and then how these combine in a rigourous manner to form basic prototypes of motion. From a small number of prototypical movements later prototype motions can be developed or elaborated. These prototypes provide a coherent but flexible system of prototypical motions in terms of which commonsense reasoning interprets movements and their causes.

Keywords: formalisation, commonsense theory, motion, primitive actions, basic schemes, rules, prototypes of motion

Part 1 Introduction and overview

1.1 Introduction

Ever since McCarthy's seminal paper "Programs with common sense" [18] it has been a goal of Artificial Intelligence to formalise aspects of common sense reasoning about the world. The subject, often called "Naive Physics", has made considerable progress, much of it through the work of Hayes [12,13]. This work may, however, be criticised on two general counts. The first is that it has tended to rely too much on the researchers' intuitions and too little on empirical evidence drawn from developmental psychology and work on children's and adults' conceptions of the world (though Hayes is clearly aware of what can be taken from Piaget, for example). The second is that the nature, as opposed to the content, of the formalisations proposed, has owed too much to forms which are computationally available (mainly logic and production systems) and too little to what may be suggested as necessary by an analysis of

NATO ASI Series, Vol. F 86
Intelligent Learning Environments
and Knowledge Acquisition in Physics
Edited by A. Tiberghien and H. Mandl

developmental and other psychological theories, notably the probable nature of mental models [8,15], where current theories point (as does Piaget's theory) to the importance in reasoning of the mental manipulation of imagined objects. In this paper, we wish to propose elements and structures which seem to us to be necessary components of a formal account of the development of common sense reasoning about motion. Our proposals are not cast in a fully formal language, but point towards what an appropriate formal language may be required to do.

Previously, one of us [23] developed that part of Hayes Naive Physics Manifesto [12] to do with motion, proposing an outline of a common sense theory of motion in terms of prototypical motions, which has been the subject of a number of empirical tests [2,3,27] Then we attempted to trace how such ideas about motion may be built up from the most primitive elements available to the infant. We suggested how these are later elaborated and combined to give a coherent but flexible system of prototypical motions which common sense reasoning uses to interpret movements and their causes. In the present paper we attempt to analyse these prototypes in terms of combinations of basic schemes and rules.

1.2 The empirical and formal base

We take as a starting point the Piagetian description of the early construction of reality, to be re-interpreted and added to where it seems necessary. This account starts before the child has developed notions of objects, causes, spaces or time. It takes from Piaget the idea that in the beginning there is just action and movement. Action is known kinaesthetically, movement is known visually. Action is volitional, of the person; movement is independent of the person. Thus children's conceptions of motion can be seen as originating through their actions on and experience of the physical world, very early in their development. Later such actions will be interiorised, leading to categories of adult reasoning which ultimately derive from this early cognitive development.

Much empirical work in Science Education has focused on pupils' and students' ideas of motion, because of their importance to physics teaching. McDermott [21], reviewing the area, showed that this work revealed that not only are pupils' and students' ideas about motion very different from those which they are taught, but also that such ideas are robust and resistant to alteration by teaching. Thus Viennot [25] characterised physics undergraduates' reasoning about dynamics in terms of impetus-like beliefs. McCloskey, Caramazza & Green [19] proposed the notion of "naive beliefs" to account for their subjects' ideas about motion. McCloskey [20] went further to suggest a parallelism between his subjects' ideas and those of the early impetus theorists, in that both see objects as kept moving by internal forces and not by external ones, by a force being impressed *in* the object when it is set moving. Champagne,

Kopfer & Anderson [5], however, interpreted students' beliefs about motion as being Aristotelian in nature, in that a force is required to keep an object moving. The difference is whether forces are seen as internal or external. But from the formal point of view, all these accounts are organised as *theories:* as a structured set of propositions about the world.

By contrast, diSessa [6] denies that common sense reasoning is best regarded as a theory about the world. He proposes that people's intuitive physics is made up of a large number of fragments which he calls phenomenological primitives. These he describes as "simple abstractions from common experiences that are taken as relatively primitive in the sense that they generally need no explanation". Such fragments are called upon by common sense reasoning in an essentially fluid and ad hoc fashion, relying on finding a match between their patterns and what is observed or anticipated.

We shall follow a similar approach to that of diSessa in defining prototypical motions, but unlike diSessa we regard them as forming a system and as having a developmental history. But we will base their construction on combinations of even more primitive elements, ultimately on actions and simple behaviours of entities.

1.3 Actions, schemes, rules and prototypes

Our most fundamental primitives will be *primitive actions.* These are restricted to actions available to the child at or soon after birth, as yet uncoordinated and not incorporated into complex sets of actions directed towards any goal. Actions combine into *schemes.* A scheme is a set of actions which constitutes part of a way of acting on the world, or, arising out of actions, a partial way of looking at the world.

It is convenient, though perhaps not necessary, to add two further categories. We give the name *rules* to schemes which express general constraints on, or expectations of, the way things behave or the way things should be conceived. We take these rules to be widely used and stable over a long period, generally through adulthood. Rules are the nearest our account comes to being propositional in character. We give the name *prototypes* to complexes of actions, schemes and rules which together form a pattern of behaviour which can be used to recognise, interpret and make predictions about motions. The distinction between scheme and prototype is fuzzy: the latter term is used when the complex it points to makes contact with recognisable and common adult ways of understanding things, generally reflected in language.

The main way in which complexes are built out of simpler elements is by simple *superposition.* In some cases, on superposition certain features are also *deleted.* The other mechanism we invoke is *substitution,* in which one type of object is allowed to play the role of another.

The early prototypes are formed by combining new schemes or rules with previously established schemes, rules or primitive actions. The majority of primitive actions, schemes and rules are proper to the earliest levels of development. Later prototypes develop by differentiation and generalisation of earlier ones, together with a few new schemes and rules.

1.4 Levels and their prototypes

The discussion in Part 2 is divided into accounts of five levels of development, covering roughly the first two years of life, during which period we take the essentials of common sense reasoning about motion to become established.

It may help the reader follow the detailed discussion to see the overview of the prototypes given below. We distinguish in all fifteen prototypical motions; however of these the first six are fundamental, with the remaining nine derived from them by elaboration or by differentiation. The six fundamental prototypes are found in levels 1 to 3, covering the first twelve months of life. They are more schematic in character than the later prototypes, which come closer to being patterns of motion recognisable in everyday adult thought and language, and can more nearly be used to account for motion in general as opposed to accounting for the results of the child's own actions.

Table 1 shows the prototypes, against the levels at which they develop.

Fundamental prototypes	
Level 1 (0 - 4 months)	***put***
Level 2 (4 - 9 months):	***fall***
Level 3 (9 - 12 months)	***carry***
	move-self
	push/pull
	set-moving
Later prototypes	
Level 4 (12 - 18 months):	***walk/run***
	non-living walk/run
	lift
	roll / slide
	throw
	hit/kick
Level 5 (18 months....)	***jump***
	fly
	float

Table 1 Prototypes and levels

Part 2 The development of ideas about motion

2.1 Level 1 (0 - 4 months)
Beginnings : the most primitive actions, schemes and rules

For Piaget, action and movement are fundamental. The child knows its actions through feeling them, and knows movement through seeing it. The child comes to choose to act, but cannot choose the movements it sees. This leads to a distinction between 'near' and 'far' space: in near space we can act whilst in far space things happen independently of us. From these materials the child constructs the Kantian categories of objects, space, time and cause. Figure 1 suggests schematically certain aspects of this construction. It can be seen in terms of two triplets. The first is action, cause and time (the last in the sense of before and after) in the upper part of Figure 1. the complementary triplet is movement, object and space (the last in the sense of regions where movement happens against a more stable visual field). These are the beginnings of the more constant and stable furnishings of the child's world.

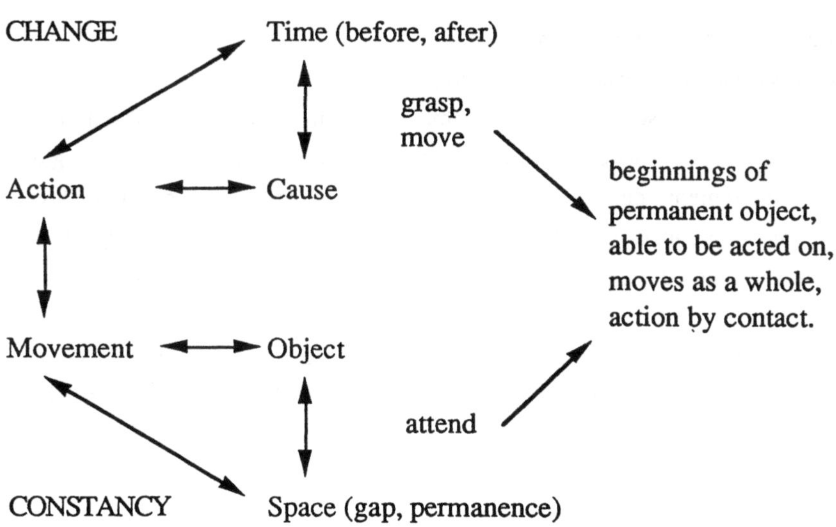

Figure 1 Origins of the fundamental categories.

With the combining of these two triplets, through actions on objects, the first set of primitive actions is established at this level (0 - 4 months).

Primitive actions:

1a *hold, move limb, let go, notice far movement*

The child is born with a grasping reflex, and noticing movement in the visual field. Grasping things, coordinating grasping and ungrasping so as to hold or let go, moving the limbs and attending to movements all play a role in the beginnings of the construction of objects, and later of their permanance. These primitive actions give rise to two basic action schemes:

Schemes:

1a *change place of object*

made up of *hold, move limb, let go*

1b *object moving on own*

made up of *notice far movement*

and a third scheme which constitutes a first way of looking at the world (one might almost call it a rule):

1c *near space: objects that can be touched*

far space: objects that can be seen but not touched

Underlying the first basic action scheme are two fundamental rules of motion. We might go so far as to say that they are the laws of motion for common sense.

Rules:

1a *the part moves with the whole*

1b *actions cause movement through contact*

These rules arise because object and movement are intimately related. The defining aspect of objects is that entities whose parts move together as a whole are seen as single objects. Thus different movements define different objects. The primitive actions of level 1 also contribute to thinking about cause. With the basic scheme *change place of object* causes are seen as actions on things through direct contact. Contact is thus built in to the very notion of cause.

A first prototype of motion can now be constructed from the two action schemes above, and the two laws of motion. We will call it *put*. We use it when as adults we reach for a cup of tea, or pass the sugar. We are *changing the place of an object.*, which is moved by being made to *move-with* the action , the object being made part of the person whilst it is grasped.

Prototype:

1a put

= Scheme 1a: *change place of object*

\+ Rule 1a: *the part moves with whole*

\+ Rule 1b: *actions cause movement through contact*

\+ Scheme 1c: *near space*

Major constructions at this level mainly take place in *near space* as with the prototype *put*. However the primitive action of *notice movement*, which gives rise to basic scheme 1b where there are *objects moving on their own, changing their position*, relates to *far space*.

2.2 Level 2 (4 - 9 months): Emerging schemes and rules

The characteristic of this level is the wide range of developing schemes and rules to do with effort and movement, with support and falling, and so with space, movement in space, and limits to movement in space. Only one new prototypical motion develops, that of *falling*.

Effort and movement

When movements of objects are attempted the child finds that many of them resist movement. Effort is felt as needed to make a movement. This gives rise to a new primitive action: *make effort*.

Primitive action:

2a *make effort* = *move hand* + *quasi-intentionality* + *resistance*

The action *make effort* combines elements of the kinaesthetic awareness of making a movement, with the beginnings of purpose or intention and of noticing the resistance of objects. Now, when this new primitive action is applied to the earlier basic action scheme(1a) to do with changing places of objects, a new basic scheme emerges:

Scheme:

2a *make effort to move object*

= action 2a: *make effort*

\+ scheme 1a: *change place of object*

From the application of this basic scheme a new rule is derived:

Rule:

2a *effort causes movement*

It is here that we see the beginnings of intentionality and causation. One might almost say, fancifully, that if there were no resistance to movement there would be no intention and no causality.

Support and falling

When the level 1 primitive actions of *grasping, holding* and *letting go* are applied to objects, objects do not always stay where they are when let go. The support scheme emerges:

Scheme:

2b *support* = *hold* or *on something*

When an object is held or when it is resting on something then it is supported, so that the *support* scheme is an extension of the primitive action *hold,* combined with spatial relations discussed below. An interesting new effect can be produced, that of letting things drop, leading to the rule that will account for these effects:

Rule:

2b *objects without support fall*

There are now enough elements in the system to construct the second prototype motion, that of *falling:*

Prototype:

2a ***fall***

= primitive actions (1a) *hold, let go*

\+ scheme (1b) *object moving on own*

\+ scheme (2b) *support*

\+ rule (2b) *objects without support fall*

This new prototype is the beginning of a link between the two previously independent schemes of level 1 (la and lb) in which either the child acts to change the place of the object or the object is seen moving on its own. Now the child can create motion of an object on its own, not by making an effort on it but by releasing it.

Space and movement

With these new schemes to do with making efforts to move objects and with the emergence of the prototype *fall,* the spatial framework and its contents develop rapidly. But first a new rule is needed about where movements can happen:

Rule:

2c *motion only takes place in empty space, in the gaps between barriers*

So there can now be added two new spatial dimensions, those of *something underneath* or *nothing underneath,* from which the spatial schemes of *above* and *below,* and *up* and *down* are derived, as well as being *on* or moving *along* a surface. Spatial relations and the scheme *support* are inextricably linked. The existence of obstacles in near space also enforces that movements happen only in gaps; at first those in front of obstacles and later, invisibly, in gaps behind obstacles. Thus two other dimensions of topological space come into being: *in front of* and *at the back of*, the second being later extended to *behind something*.

Scheme:

2c *space = above, below, up, down, on, along, in front of, behind*

Space has an absolute lower limit, namely the ground, which always gives support and cannot move. Thus the following two rules now come into existence:

Rules:

2d *the ground gives support to anything*

2e *the ground is fixed*

Space, effort and movement

Some objects when grasped will not move, so that the failure to move something fixed is noticed. Thus an aspect of space, often defining its boundaries, is to do with *fixed things* such as walls and large objects, and already includes the ground. Thus we have the scheme *fixed* :

Scheme:

2d *fixed = fail to move object with effort*

The application of this scheme generates its complementary new rule

Rule:

2f *fixed things cannot move*

It follows from all this that space itself is passive and inactive; that it does not move things or support things, except the ground which supports everything without itself being supported. Nor of course can space itself move. Thus, progressively, starting from the initial noticing of movement against a relatively stable visual field, through the gradual construction of the object, and then through actions on objects, the world comes to have a fixed framework within which motion takes place.

2.3 Level 3 (9 - 12 months): New prototypes

Four prototypes which are to serve as the basis for the elaboration of all the further prototypes which develop after this level are now constructed: *move self, carry, push/pull, set moving on own*. These, and other new rules and schemes which develop at this level, are linked to and contribute to the gradual development of intentionality, whose seeds lie earlier in the scheme *make effort to move object*.

Despite the relation of intention to effort and movement, intentionality encompasses a much wider range of interactions than these, so that describing the development of intentionality would go well beyond the scope of the present arguments. We shall therefore feel free to mention intention as an ingredient of our schemes without attempting any more detailed analysis.

Move Self

The child now begins to be able to move itself from place to place, thus changing what is near and far space, and beginning to be itself an entity moving in space. Thus we combine the primitive action *make effort* with the far space scheme *move on own*, to obtain:

Scheme:

3a *make effort to move self*

= primitive action (2a) *make effort*

+ scheme (1b) *move on own*

Such movement is intentional, and is subject to continual adjustment and change; that is, is *controlled.* Intention and control are experienced as effort. Thus we get two new rules:

Rules:

3a *things which move themselves provide their own effort*

3b *movement is controlled by effort*

Rule 3b can be seen as an extension of rule 2a *effort causes movement.* All the elements for the new prototype of *move self* now exist:

Prototype:

3a ***move*** *self*

= scheme (3a) *make effort to move self*

+ rule (3a)*things which move themselves provide their own effort*

+ rule (3b) *movement is controlled by effort*

+ *intention*

Such a prototype will be absolutely fundamental, since it allows the possibility of distinguishing those things, mainly people, which move of their own accord, from those which move because an entity acts on them. Here again is the seed of another scheme, *living things,* and the distinction *living/non-living*, which like intention, go well beyond the limits of the present analysis. Because living things move themselves, and things which move themselves provide their own effort (rule 3a), living things, including the self, are seen as a source of effort. Effort and intention become closely tied. These distinctions will take time to develop, but lead to a difference between movement whose cause is effort from within (intention) and movement whose cause is effort from without (by an intentional agent). Thus the origin of non-living movement is conceived of as an action of a living thing on a non-living thing:

Rule:

3c *living things can move themselves*
non-living things are made to move

Carry

When the child can move in space, it can move objects with it. We previously noted the fundamental character of the rule (1a) *the part moves with the whole.* This now extends from defining objects, to being the basis of a new kind of movement, namely being *carried.* Things held by or on, and so supported by, a moving object travel with it.

Prototype:

3b ***carry***

= prototype (1a) *put*
\+ prototype (3a) *move self*
\+ scheme (2b) *support*

Intention is built in via the prototype *move self.* The prototype *carry* can be seen as a development of the scheme *change place of object* in which the object *moved with* the action of the child, but now with the carrier itself moving as an entity. First constructed through the child's own movements, and through the experience of itself being carried, later the prototype *carry* (unknown to physics) will become a fundamental mode of motion, being extended to cars, trains and aeroplanes.

Push/ Pull

The child can now also move objects more deliberately, at first pushing or pulling them near itself, and then pushing or pulling while moving. Such objects are supported by the ground, not by the child. So we get the prototype *push/pull* , which represents such motions as pushing a pram or pulling a cart in which an agent controls the movement of something, normally supported by the ground:

Prototype:

3c ***push/pull***

= scheme (2a)*make effort to move object*
\+ rule (3b)*movement is controlled by effort*
\+ scheme (2b) *support*
\+ *intention*

As we have noted, *move self, carry* and *push/pull* are all linked to developing ideas of *control,* themselves related to *intention.* Effort becomes directed not just to making movements but to making them happen as desired. Rule 1b *actions cause movement through contact* ensures that control is seen as exercised through contact, that is, as deriving ultimately from the primitive action *hold.*

Set moving on own.

Until now, an object would only move on its own when left to fall. A new scheme is now constructed so that objects can be set moving on their own. This arises from a combination of an earlier scheme from level 2, where *making an effort* was projected onto objects, together with the primitive action *letting go* from level 1 and also of noticing in far space *objects moving on their own , changing position.* In other words, an object in near space can be made to move on its own in a way similar to an object in far space if an effort is made on it and the object is then released. We use an arrow symbol in place of the + symbol, to indicate that the action follows in sequence, not as a superposition.

Scheme:

3b *make object move on own*

= scheme (2a) *make effort to move object*

-> primitive action (1a) *let go*

\+ scheme (1b) *object moving on own*

Underlying this scheme is the rule:

Rule:

3d *effort is given to an object made to move on own*

This effort can not, however, control the motion of the object after it is set moving, because control is exerted through contact, and what is fundamental to objects moving on their own is the intentional releasing of contact (*let go*), so we get the rule:

Rule:

3e *an object moving on its own is not controlled*

These elements now combine in another new prototype, that of objects set moving on their own:

Prototype:

3d set moving on own

= scheme (3b) *make object move on own*

\+ rule (3d) *effort is given to an object made to move on own*

\+ rule (3e) *no control*

\+ *intention*

2.4 Level 4 (12 -18 months): Emergence of new prototypes; differentiation and generalisation

The prototypes which develop at this level are more recognisable as existing in adult thought and language. They are all specialisations or extensions of the previous prototypes.

Walk/run

The distinction between *living / non living* things was based on living things being the source of their own movement, but it now has to be accepted that non-living things, such as clockwork toys, cars and buses, seem also to share this power. That a living thing can walk or run is something the child knows as a felt and lived experience. It also appears that other autonomously moving objects can do something similar, though just how they do it is unclear. If a non-living thing does move of its own accord, it must at least have its own internal supply of effort. However such a distinction will remain problematic for children for many years, and autonomous movement will often remain associated with living things. We now distinguish two related prototypes:

Prototypes:

4a *walk/run*

= prototype (3b) *move self*

\+ rule (3c)*living things can move themselves*

\+ rule (2d)*ground gives support to anything*

4b non-living *walk/run*

= prototype (4a)*walk / run*

\- rule (3c)*non-living things are made to move*

\+ rule (3a)*things which move themselves provide their own effort*

We include the rule about support from the ground here because this motion, unlike some which follow, is conceived of as taking place on the ground, though this assumption is so much taken for granted as to be almost invisible.

Lift

As objects are moved, by pushing, pulling or carrying, the special nature of the vertical direction becomes more salient. Objects tend to resist being moved upwards. They are felt as heavy. Effort is felt as needed to give them support.

Scheme:

4a *heavy*

= scheme (2b) *support*

\+ primitive action (2a) *make effort*

\+ scheme (2c) *spatial direction - up*

Rule:

4a *effort is needed to support a heavy object*

Prototype:

4c *lift*

= prototype (3c) *push/pull* and/or prototype (3b) *carry*

\+ scheme (4a) *heavy*

\+ rule (4a) *effort is needed to support a heavy object*

Roll, throw and hit

The prototype 3d *set moving on own* can now become differentiated. The two main new prototypes are *roll/slide* and *throw*. They differ in that the first is for things set moving on the ground, while the second is for things set moving in the air. Both involve the effort used to control direction which is part of *push/pull.*

A thing can be rolled or slid along the ground by first *pushing* it and then by *setting it moving on its own.* The two prototypes combine in temporal succession: *push* initiates the action and *set moving on own,* involving *let go* , completes it. The movement is at first controlled, and then not. Again, an arrow symbol indicates sequence.

Prototype:

4d ***roll or slide***

= prototype (3c)*push/pull*

-> prototype (3d)*set moving on own*

\+ scheme (2b)*spatial direction along*

\+ rule (2e) *ground gives support to anything*

To throw something is to hold and support it, then to both push it with the movement of the arm and give it the movement of the arm - a combination of *push* and *carry.* The *push* involves effort given to the object and effort used to control direction. Then it must be *let go,* so being *set moving on its own.* But something thrown is not then supported, so it will next start to fall. As with *roll/slide,* there is a temporal succession of actions.

Prototype:

4e *throw*

= prototype (3b)*carry*

-> prototype (3c) *push*

-> prototype (3d) *set moving on own*

\+ scheme (2c) *spatial direction up*

-> prototype (2a) *fall*

Throw may involve prototype 4c *lift* as well as *push*. In addition to *roll/slide* and *throw* we also need a prototype very like *set moving on own,* to represent actions like hitting or kicking a thing (which may or may not be supported by the ground after it is hit).

Prototype:

4f *hit or kick*

= prototype (3d) *set moving on own*

\+ scheme (2c) *spatial direction across and/or up*

->

either rule (2d)*ground gives support to anything*

or prototype (2a) *fall*

The difficulty of control with *roll/slide, throw and hit* after the object is set moving on its own forms the basis of many games. In some (e.g. darts or bowls) the element of initial control, as in *throw* or *roll / slide* , is retained. One might say that all such games consist of making *change place of object* difficult in some way.

Stopping motion

In all prototypes except falling, effort is involved in setting an object moving, either from itself or from an agent. The two ideas can be brought together by supposing that any moving object contains effort, so that non-living objects set moving from outside are seen as taking on the effort provided when they were set moving (rule 3d).

We can now consider how things stop moving. The necessary elements are already there, in the basic scheme of space as composed of gaps between fixed barriers. Falling was always stopped by the ground. Likewise, rolling, sliding or thrown motions often stop at fixed walls or obstacles. The earlier rule that motion only takes place in empty spaces can now be elaborated to include movements stopped by barriers.

Not all such motions are stopped in this way, however. A ball set rolling may just stop. If this motion is seen as kept moving by effort inside the object, it can be understood as stopping because the effort has been used up. So two new rules emerge:

Rules:

4b *Movement is stopped by barriers*

4c *Non-living objects stop if their effort is used up.*

The idea that effort is used up comes from the fact that a non-living object does not have its own source of effort (rule 3c) but only the effort given to it when it was set moving (rule 3d). If an object is thrown upwards, the effort given to it may also be seen as being used up by supporting it (rule 4a), so that it soon begins to fall.

Generalisation

A number of schemes and rules are now developed that will be used in generalising previous prototypes. Only now will the main prototypes and schemes so far established be applied to account for motion in general as opposed to accounting for results of the person's own actions. So far, all action is seen as originating in the self or in the movement of a living thing. At this level however, the self begins to be able to be seen form 'outside', as a moving object like other moving objects, and not as unique:

Scheme:

4b *self as a moving object like other moving objects*

This scheme arises from the gradual decentering of the self. The earlier distinction between living and non-living things is maintained, with the notion that living things can intentionally move themselves whereas non-living objects can not, being retained. But the distinction now begins to be used more flexibly, imagining non-living things playing the role of living things, and vice-versa. Thus there emerge two rules crucial to the generalisation of prototypes of motion:

Rules:

4d *other objects can be substituted for self in prototypes*

4e *self can be substituted for other objects in prototypes*

Using scheme 4b and rules 4d and 4e, the construction of motion takes on a new dimension. It is now possible to project onto objects the possibility of acting like the self, for example in *throw*, for example if an object is 'thrown' by a gun. Besides projection, there is also introjection of the idea of an object and its movements, so that the self can imagine being an object among other objects.

Another new scheme is do with the notion of *agent*.. With the *push/pull* prototype an object can be brought closer or pushed further away in near space. But it can be observed that *living things* can also push objects away or bring them closer in near or far space. The object moves but is clearly under control of the *agent* who is moving it intentionally. In other words, another person is seen as the *source of effort* and this combines with the ideas of *control* and *intention*. Thus the distinction between *living* and *non-living things* is important for the establishment of this scheme:

Scheme:

4c *agent = thing making effort to move, support or control objects*

Scheme 4b thus derives from the combination of many earlier schemes and rules as well as scheme 4a. Implied by the new scheme is a rule:

Rule:

4f *an agent can give effort to an object made to move on own*

At this point, where there has been sufficient decentering for the self to become an object amongst other objects, and for other objects to become agents equivalent to the self, the prototype *carry* (3b) can be extended to a wider range of carriers. The self can both *carry* and be *carried* - as can other objects. Because a carried object moves as part of a whole (rule 1a), a new rule about inheritance develops which specifies that a carried thing moves because the carrier moves:

Rule:

4g *a carried object inherits the prototypical motion of its carrier*

This rule underlies our perception that the passengers in an aeroplane fly because the plane flies. When the self has become a moving object amongst other objects, and can see other objects act like agents of motion, existing schemes and prototypes can be applied to the actions of objects on other objects, and on the self. For example, children are subject to the impact of objects on themselves. A moving object can *hit* the self, who feels the impact, making it yet more reasonable to think of effort in the moving object. Similarly, children will construct the idea that they too are objects which can be *pushed* or *pulled*.

2.5 Level 5 (18 months on): Further elaboration

From this level onwards no fundamentally new primitives or basic schemes are needed, and only two new rules develop, as the prototype of *run* from level 4 is elaborated into two new prototypes. Young children climb and jump over obstacles, and begin to see themselves as objects negotiating barriers in space. From this activity, by analogy with the movement of *throwing* an object, the coordination *run + use effort + throw self* can be the basis for the new prototype *jump*.

Prototype

5a *jump*

= prototype (4a)*run*

+*throw self*

= [prototype (4e) *throw*

+ scheme (4b)*self as a moving object*

+ rule (4e) *self can be substituted for other objects in prototypes*]

For young children *far space* expands to encompass objects such as birds which have no visible support but do not fall, instead they move by themselves. The bird seems to *hold itself up* by flapping its wings. Previously, *effort* had *made self move* or *made objects move*, but was also used to *provide support* (rule 4a). So for objects like birds rule 4a is generalised:

Rule:

5a *effort can provide support*

A new prototype *fly* which combines using effort to move and to support is now possible:

Prototype

5b *fly*

= prototype (5a) *jump*

+ rule (5a) *effort can provide support*

This solution means that, since it derives from *make effort to move self*, a flying object must necessarily have its own source of effort, assumed for living things and required for aeroplanes.

There are however objects such as balloons, feathers, tissue paper that have no visible support, nor any source of effort, but which nevertheless do not fall immediately. Sense can be made of this by the projection of the *carry* prototype. One way of understanding the floating of light objects is to see them as carried, and thus held up, by the air. If this is so, it makes sense that they move with the air, not independently. In a similar manner some objects can be carried on water and not sink. Rather obviously, the support offered by water or the air is often at best partial, so a new rule can be constructed:

Rule:

5b *support can be partial*

With this elaboration of the prototype *carry* together with the new rule about *partial support* the new prototype of *float* can develop.

Prototype

5c *float*

= prototype (3b) *carry* *(by air or by water)*

\+ rule (5b) *partial support*

Clearly some motions can be seen as a mixture of falling and floating. In any case, these new elaborations have a less fundamental character than earlier constructions. It is possible to use them flexibly, without strong commitment. Thus a dropped scrap of tissue paper may be seen as falling or as floating down; or a bird as at one moment flying on its own and at the next as carried by the air.

Part 3 Conclusions

3.1 Prototypes and expectations

The point of prototypes, despite the complex developmental history traced in this paper, is to provide a simple, efficient and rapid means of recognising and making sense of motions, in such a way that inferences about what is going on are immediate and 'obvious' - indeed generally barely conscious. This process can best be illustrated by some simple everyday examples.

1. A person goes to the bus stop and catches a bus

The trip to the bus-stop belongs to the prototype *walk*. It follows at once that the person makes the necessary effort, can control the motion, and won't fall because supported by the ground. When on the bus, the person is *carried* by the bus. We know without further reflection that the person, being inside the bus, will behave like a part of a whole and so will move with the bus (*non-living walk-run*); that the bus, not being alive, needs an engine to provide effort; and that neither bus nor person will fall because the bus is supported by the ground and the person by the bus.

2 A ball game

A player can *run, carrying* the ball. The ball moves because the player does, and is supported by the player. Then the player may *throw* or *kick* the ball. To do it the player must provide *effort* on the ball and must *let go*. If the ball is thrown in the air, it is *not supported* and will ultimately *fall* after *moving on its own*. Kicked along the ground, it is *supported* and may, after continuing to *move on its own* because of the *effort* given to it, come to rest having used up this effort.

3. TV pictures of a space-man

Pictures of a man in space are not so immediately interpretable. There is nothing underneath the space-man - *no support* - and yet he does not *fall*. One possible prototype might be *float*. But this would mean that the space-man is *carried* by *space*. This may itself make a problem: things move *through* space, which is itself empty and therefore inactive. So space may start to be thought of as not really empty.

The last example reveals that the process of interpretation through prototypes is essentially *abductive:* that is, trying to find a way of imagining the world for which what seems to be happening makes evident sense.

3.2 Status of the analysis

There are a number of difficulties in the way of arriving at any estimate of the status or value of the foregoing analysis.

One difficulty is that it is not cast in any fully formalized language. This makes it difficult to check the analysis for internal consistency or for redundancy. We have, however, tried to express much of it as formally as seemed possible, in order that the logic of the structure be as clear as possible. There is probably some redundancy, especially in the later rules.

Another difficulty is just the reverse, that the analysis *is* cast in a rather formal language. Formal systems, including computational ones, do not construct themselves from their experience. They do not occupy an experiential world in which their actions and experience are primary, prior to and constitutive of any system of interpretation. For just this reason there will be those who find our analysis repulsive. Our view is that if the understanding of motion does have a structure built on action and experience, it is important to try to elucidate that structure in whatever way seems possible. And secondly, that attempting to do so may usefully illuminate the difficulties involved in formalising that kind of structure.

We do believe that we are fundamentally correct in building the structures mainly out of superpositions of behaviours of things. A new behaviour, usually associated with a particular kind of thing, is understood as a combination of the behaviours of previously understood things. We also believe we are correct in giving effort and support a fundamental role, together with the insistence on the relationship of movement and the construction of objects, which leads via *the part moves with the whole* to the prototype *carry*. These last aspects do not seem to have received the attention they deserve, in either the Artificial Intelligence and the psychological literature.

Ultimately, we regard the present attempt as worth making, not because it is likely to be a final or definitive account of the development of reasoning about motion, but because, by existing as a proposed account, it raises a number of questions. Amongst them are the question as to whether this much complexity is really necessary; the countervailing question as to whether reasoning is not altogether too fluid to be captured by such an analysis; questions about the relation to language and context; and questions about the use of formal methods in this kind of area of inquiry.

One final prejudice should be set on record. It may be that the focus on reasoning about motion is not as limited as it may seem. A not inconsiderable part of human reasoning in general seems to be based on metaphors of effort, movement and support. An understanding of our ideas about motion may be one window onto more general kinds of reasoning.

References

1. Andersson, B. : 'The experiential gestalt of causation: a common core to pupils' preconceptions in science', European Journal of Science Education, Vol 8, No 2, pp 155-171, 1986.
2. Bliss, J. Ogborn, J. Whitelock, D. : 'Secondary school pupils' common sense theories of motion', International Journal of Science Education , Vol 11, No 3, 261-272, 1989.
3. Bliss, J. Ogborn, J. : 'A common sense theory of motion: issues of theory and methodology examined through a pilot study', in P. Black & A. Lucas (Eds) Children's informal ideas about science (pp.158-172), London: Croom Helm, Routledge, 1990.
4. Bower , T. G. R. : 'The development of object permanence: some studies of existence constancy', Perception and Psychophysics, Vol. 2, 74-6, 1967.
5. Champagne, A.B. Kopfer, L.E. Anderson, J.H. : 'Factors influencing the learning of classical mechanics', American Journal of Physics, 48, 1074-1079, 1980.
6. diSessa, A. : 'Knowledge in Pieces', in G. Forman & P. B. Pufall (Eds) Constructivism in the Computer Age , (pp.49-70), Hillsdale, New Jersey: Lawrence Erlbaum Associates, 1989.
7. Flavell, J. : The developmental psychology of Jean Piaget . Princeton, New Jersey: D. Van Nostrand Co. Inc., 1963.
8. Gentner, D., Stevens, A.L. : Mental Models, New Jersey: Lawrence Erlbaum, 1983.
9. Guidoni, P. Arca, M . Mazzoli, P. : 'Structures of understanding at the roots of science education, Part I: Experience, language and knowledge', European Journal of Science Education Vol 5, No 4, 367-375, 1983.
10. Guidoni , P. Arca, M. Mazzoli, P. : 'Structures of understanding at the roots of science education, Part II: Meanings for formalisation', European Journal of Science Education Vol 6, No 4, 311-319, 1984.
11. Guidoni, P. : 'On Natural Thinking', European Journal of Science Education Vol 7, No 2, 133-140, 1985.
12. Hayes, P. : 'The naive physics manifesto', in D. Mitchie (Ed) Expert Systems in the Microelectronic Age , Edinburgh, Edinburgh University Press, 1979.
13. Hayes, P. : 'The second naive physics manifesto', In Hobbs, J.R., Moore, R.C. Formal theories of the common sense world, New Jersey: Ablex, 1985.
14. Hobbs, J.R., Moore, R.C. : Formal theories of the common sense world, New Jersey: Ablex, 1985.
15. Johnson-Laird, P.N. : Mental models, Cambridge: Cambridge University Press, 1983.
16. Leslie, A. M. : The necessity of illusion: perception and thought in infancy. In L. Weiskrantz (Ed.) Thought without Language , (pp.185-210) Oxford: Oxford University Press, 1988.
17. Lakoff, G. Johnson, M. : Metaphors we live by, Chicago: University of Chicago Press, 1980.
18. McCarthy, J. : 'Programs with common sense', in Minsky M (Ed) Semantic information processing ,

Cambridge, Mass.: MIT Press, 1968.
19. McCloskey, M. Caramazza, A. & Green, B.: 'Curvilinear motion in the absence of external forces: Naive beliefs about the motion of objects', Science, 210, 1139-1141, 1980.
20. McCloskey M.: 'Naive Theories of Motion', In Gentner D., Stevens, A.L. (Eds) Mental Models, (pp. 299-324), Hillsdale, New Jersey: Lawrence Erlbaum Associates, 1983.
21. McDermott, L.: 'Critical review of review of research in the domain of mechanics', Recherche en Didactique de la Physique, les Actes du Premier Atelier international, La Londes les Maures 1983, (pp.137-182) Paris: CNRS, 1984.
22. Minsky M (Ed): Semantic information processing, Cambridge, Mass.: MIT Press, 1968.
23. Ogborn, J.: 'Understanding students' understandings: an example from dynamics', European Journal of Science Education, Vol 7, No 2, 141-162, 1985.
24. Piaget, J.: La Construction du Réel chez l'enfant, Neuchâtel: Delachaux & Niestlé, 1937.
25. Viennot, L.: 'Spontaneous reasoning in elementary dynamics', European Journal of Science Education, Vol.1, No..2, 205-221, 1979.
25. Viennot, L.: 'Analysing students' reasoning in science: A pragmatic view of theoretical problems', European Journal of Science Education, Vol 7, 151-162, 1985.
27. Whitelock, D.: Common sense understandings of causes of motion, unpublished PhD thesis, Institute of Education, University of London, 1990.
28. Wiser M, Carey S: 'When Heat and Temperature Were One', In Gentner, D., Stevens, A.L. (Eds) Mental Models (pp. 267-298), Hillsdale, New Jersey: Lawrence Erlbaum Associates, 1983.

Appendix : Summary of schemes, rules and prototypes

Level 1 (0 - 4 months)

Primitive actions:
1a *hold, move limb, let go, notice far movement*

Schemes:
1a *change place of object*
made up of *hold, move limb, let go*
1b *object moving on own*
made up of *notice far movement*
1c *near space: objects that can be touched*
far space: objects that can be seen but not touched

Rules:
1a *the part moves with the whole*
1b *actions cause movement through contact*

Prototype:
1a *put*
= Scheme 1a: *change place of object*
+ Rule 1a*: the part moves with whole*
+ Rule 1b: *actions cause movement through contact*
+ Scheme 1c: *near space*

Level 2 (4 - 9 months)

Primitive action:
2a *make effort = move hand + quasi-intentionality + resistance*

Schemes:
2a *make effort to move object*
= action 2a: *make effort*
+ scheme 1a:*change place of object*
2b *support = hold or on something*
2c *space = above, below, up,.down, on, along, in front of, behind*
2d *fixed = fail to move object with effort*

Rules:
2a *effort causes movement*
2b *objects without support fall*
2c *motion only takes place in empty space, in the gaps between barriers*
2d *the ground gives support to anything*
2e *the ground is fixed*
2f *fixed things cannot move*

Prototype:
2a *fall*
= primitive actions (1a) *hold, let go*
+ scheme (1b) *object moving on own*
+ scheme (2b) *support*
+ rule (2b) *objects without support fall*

Level 3 (9 - 12 months)

Schemes:
3a *make effort to move self*
= primitive action (2a) *make effort*
+ scheme (1b) *move on own*
3b *make object move on own*
= scheme (2a) *make effort to move object*
-> primitive action (1a) *let go*
+ scheme (1b) *object moving on own*

Rules:

3a *things which move themselves provide their own effort*
3b *movement is controlled by effort*
3c *living things can move themselves*
non-living things are made to move
3d *effort is given to an object made to move on own*
3e *an object moving on its own is not controlled*

Prototypes:

3a *move self*
= scheme (3a) *make effort to move self*
\+ rule (3a)*things which move themselves provide their own effort*
\+ rule (3b) *movement is controlled by effort*
\+ *intention*

3b *carry*
= prototype (1a) *put*
\+ prototype (3a) *move self*
\+ scheme (2b) *support*

3c *push/pull*
= scheme (2a)*make effort to move object*
\+ rule (3b)*movement is controlled by effort*
\+ scheme (2b) *support*
\+ *intention*

3d *set moving on own*
= scheme (3b) *make object move on own*
\+ rule (3d) *effort is given to an object made to move on own*
\+ rule (3e) *no control*
\+ *intention*

Level 4 (12 - 18 months)

Schemes:

4a *heavy*
= scheme (2b) *support*
\+ primitive action (2a) *make effort*
\+ scheme (2c) *spatial direction - up*

4b *self as a moving object like other moving objects*
4c *agent = thing making effort to move, support or control objects*

Rules:

4a *effort is needed to support a heavy object*
4b *movement is stopped by barriers*
4c *non-living objects stop if their effort is used up.*
4d *other objects can be substituted for self in prototypes*
4e *self can be substituted for other objects in prototypes*
4f *an agent can give effort to an object made to move on own*
4g *a carried object inherits the prototypical motion of its carrier*

Prototypes:

4a *walk/run*
= prototype (3b) *move self*
\+ rule (3c)*living things can move themselves*
\+ rule (2d)*ground gives support to anything*

4b *non-living walk/run*
= prototype (4a)*walk / run*
\- rule (3c)*non-living things are made to move*
\+ rule (3a)*things which move themselves provide their own effort*

4c *lift*
= prototype (3c) *push/pull* and/or prototype (3b) *carry*
\+ scheme (4a) *heavy*
\+ rule (4a) *effort is needed to support a heavy object*

4d *roll or slide*

= prototype (3c)*push/pull*
-> prototype (3d)*set moving on own*
+ scheme (2b)*spatial direction along*
+ rule (2e) *ground gives support to anything*

4e *throw*

= prototype (3b)*carry*
-> prototype (3c) *push*
-> prototype (3d) *set moving on own*
+ scheme (2c) *spatial direction up*
-> prototype (2a) *fall*

4f *hit or kick*

= prototype (3d) *set moving on own*
+ scheme (2c) *spatial direction across and/or up*
->
either rule (2d)*ground gives support to anything*
or prototype (2a) *fall*

Level 5 (18 months on)

Rules:

5a *effort can provide support*
5b *support can be partial*

Prototypes:

5a *jump*

= prototype (4a)*run*
+*throw self*
= [prototype (4e) *throw*
+ scheme (4b)*self as a moving object*
+ rule (4e) *self can be substituted for other objects in prototypes*]

5b *fly*

= prototype (5a) *jump*
+ rule (5a) *effort can provide support*

5c *float*

= prototype (3b) *carry* *(by air or by water)*
+ rule (5b) *partial support*

Computerized Analysis of Students' Ability to Process Information in the Area of Basic Electricity

Karl Grob[1], Victor L. Pollak[2], and Christoph von Rhöneck[1]

[1]Pädagogische Hochschule Ludwigsburg, D-7140 Ludwigsburg, FRG
[2]University of North Carolina at Charlotte, NC 28223, USA

Abstract: A test was administered by PC, using an expert system. In the first half of the test students received various kinds of feedback specific to their answers. The feedback given was intended to influence the way students solve similar tasks in the second half of the test. Four different test versions were employed, including one version with no feedback. In the analysis, students were grouped according to the ways in which they structure knowledge. The students' preexisting knowledge structure was found to be more important for the solution of the tasks than the given feedback.

Keywords: expert system, basic electricity, students' conceptions, knowledge structure, information processing

1. Introduction

Based on research work on students' conceptions in basic electricity [3,4] a computerized test on learning difficulties was developed. The test was used to study changes in students' conceptions when the student gets feedback on his/her solution in the first part of the test and then tries to apply this information in the second part. The information the student receives directly after selecting an answer from a multiple choice grid is given in four versions: (1) no information, (2) right/wrong information, (3) short information, and (4) detailed information on the correct solution and on underlying concepts.

The test was presented some months after instruction to 15 to 17 year-old students in a Realschule (a kind of technical school, N = 120, grade 9), in a Gymnasium (N = 128, grade 10), in a Junior High School (N = 113, grade 9), and in a Technical Gymnasium (N = 119, grade 11) whose students are specially oriented towards science and technology. Each of the four test versions was administered to approximately 120 students. The time required to take the test was about 25 minutes. The students' solutions were recorded by the expert system (VP-

NATO ASI Series, Vol. F 86
Intelligent Learning Environments
and Knowledge Acquisition in Physics
Edited by A. Tiberghien and H. Mandl

Expert [5]) used to administer the test. The advantages of using an expert system may be seen in it's simplicity (i.e. automatic question generation, "inference engine" with backward and forward chaining, rapid execution of the knowledge base, etc.). The expert system analyses the students' responses in terms of previously identified students' conceptions [4,3]. The results are then evaluated statistically using SPSS/PC^{+}.

The test consists of 8 pairs of exercises with similar, but not identical tasks: exercise 1 corresponds to exercise 9, exercise 2 to exercise 10, etc. The pairs are related to the following well-known students' conceptions [3,4]: In exercise 1 and exercise 9 the student is asked to classify simple circuits with resistors connected in parallel and in series. These two exercises serve as introduction and are not considered in the analysis. Exercises 2 and 10 are related to semantic difficulties, such as 'differentiating between voltage and current' and 'static currents' in open wirings. Exercises 3 and 11 aim to identify the misconception that 'current is consumed'. Exercises 4 and 12 examine 'sequential reasoning'. In 'sequential reasoning' a circuit is seen in terms of 'in front of' and 'behind', the student believing that the current in a particular resistor is influenced by circuit changes 'in front of' the resistor, whereas a change 'after' or 'behind' will leave the resistor current unchanged. Exercise 5 and exercise 13 ask about voltages in a simple circuit, and in a circuit with two bulbs connected in series. The problem for the student in exercises 6 and 14 is to find voltages in circuits having two or three resistors connected in parallel. Exercises 7 and 15 focus on 'local' reasoning. Local reasoning means that current splits into two equal parts at a junction regardless of what is connected elsewhere. The last pair of exercises (8 and 16) is related to a form of reasoning called 'compensation'. Here any change in the current in one branch of a parallel connection is always compensated for by an inverse change in the current in the other branch.

The pair of exercises 6 and 14 is presented in Fig. 1. For each of the posssible answers there is a relation between the answer and the underlying student conception. These relations stem from previous investigations and from interviews. For example, solution 2 in exercise 6 and solution 3 in exercise 14 correspond to the conception that the voltage "splits up" into two equal parts at every junction point. A typical local argumentation is: "Sure, when it is 6V before the junction point, then it divides up here, and 3V go down and 3V go up." Solution 3 in exercise 6 and solution 2 in exercise 14 correspond to the conception that the voltage behaves like the current, solution 4 in exercise 6 that it behaves inversely. A typical student's comment is: "A smaller current flows through the bigger resistor." This looks as if voltage and current have been confused. The hint of the interviewer that the exercise asks for the voltages and not for the current, provoked in the cited interview the remark: "You can multiply them, V is R times i, and by multiplication you come to the same result." Obviously, the student sees only the proportionality between voltage and current, but does not distinguish the concepts. The last

Exercise 6

The following circuit has two different resistors, $R_1 = 10\,X$ and $R_2 = 20\,X$.

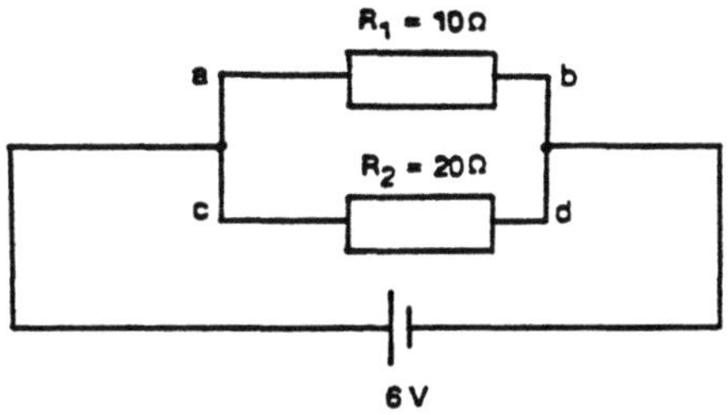

In this circuit, what is the voltage between point a and point b (V_{ab}), and between point c and point d (V_{cd})?

Select the correct answer from the following possibilities:

Solution	S_1: $V_{ab} = 6V$	$V_{cd} = 6V$	(V_corr_6)
	S_2: $V_{ab} = 3V$	$V_{cd} = 3V$	(V_loc_6)
	S_3: $V_{ab} = 4V$	$V_{cd} = 2V$	(V_as_i_6)
	S_4: $V_{ab} = 2V$	$V_{cd} = 4V$	(V_as_i_6_invers)
	S_5: $V_{ab} = 0V$	$V_{cd} = 0V$	(V_geom_6)

Exercise 14

The three bulbs in the following circuit are identical.

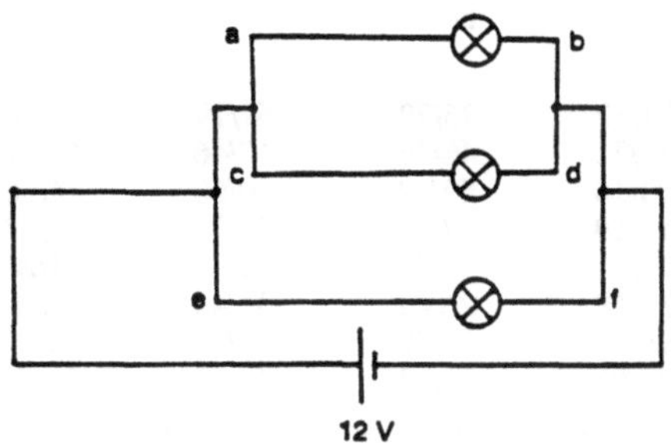

What is the voltage between point a and point b (V_{ab}), between point c and point d (V_{cd}), and between point e and point f (V_{ef})?

Select the correct answer from the following possibilities:

Solution	S_1: $V_{ab} = 12V$	$V_{cd} = 12V$	$V_{ef} = 12V$	(V_corr_14)
	S_2: $V_{ab} = 4V$	$V_{cd} = 4V$	$V_{ef} = 4V$	(V_as_i_14)
	S_3: $V_{ab} = 3V$	$V_{cd} = 3V$	$V_{ef} = 6V$	(V_loc_14)
	S_4: $V_{ab} = 0V$	$V_{cd} = 0V$	$V_{ef} = 0V$	(V_geom_14)

Figure 1: Exercises 6 and 14 with solutions and conceptions (in brackets). The conceptions are not presented to the students

solutions given in exercises 6 and 14 are attractive to students who learned the voltage concept with the help of a gravitational analogy. Remarks like "there is no height difference" show that the student does not distinguish between geometrical and electrical height.

The excerpts cited from the interviews show some typical students' conceptions. However, the remarks are often less clear. In some cases the student tries to transfer a supposedly successful strategy from one task to another. Generally the students do not mark the solutions arbitrarily, but are often uncertain about how they arrived at a solution. This uncertainty becomes apparent when the student ticks the correct solution and is then asked for his reasons.

2. First Impressions and First Results

A first impression of the results is given in Tab. 1 and Tab. 2 showing the solution frequencies of the items for different schools and feedback versions. The most striking result is that there are no statistically significant differences among the results of the four versions.

Version: Pairs of exercises	no information	right/ wrong information	short information	detailed information	mean
1/09	71/82	76/79	69/79	71/76	72/79
2/10	45/22	46/36	47/46	49/52	47/39
3/11	73/46	69/49	80/60	74/56	74/53
4/12	18/19	27/26	27/26	16/28	22/25
5/13	21/21	24/30	31/36	24/28	25/29
6/14	30/49	26/47	31/45	34/39	30/45
7/15	29/53	30/40	30/43	33/39	30/44
8/16	41/08	51/11	40/17	43/09	44/11
mean 2-8/10-16	37/31	39/34	41/39	39/36	39/35

Table 1: Solution frequencies for the different versions of information and pairs of exercises as percentages

The feedback in the first part of the test obviously has only little influence on the results in the second part of the test. On the other hand, there are statistically significant differences among the schools. As expected, the students in the Gymnasium and in the Technical Gymnasium show the best results. Also, it will be noted that the students in the Realschule and in the Junior High School actually show poorer results in the second part of the test than in the first part.

Version: Types of schools	no information	right/ wrong information	short information	detailed information	mean
Realschule, N=120	39/36	37/25	42/35	43/29	40/29
Gymnasium, N=128	35/35	35/33	36/41	34/41	35/38
Techn.Gymnasium, N=119	41/44	48/52	44/59	43/51	44/51
Junior High School, N=113	32/20	35/24	41/23	37/20	36/22
mean, N=480	37/31	39/34	41/39	39/36	39/35

Table 2: Solution frequencies for the different versions of information and types of schools as percentages (exercises 2-8/exercises 10-16)

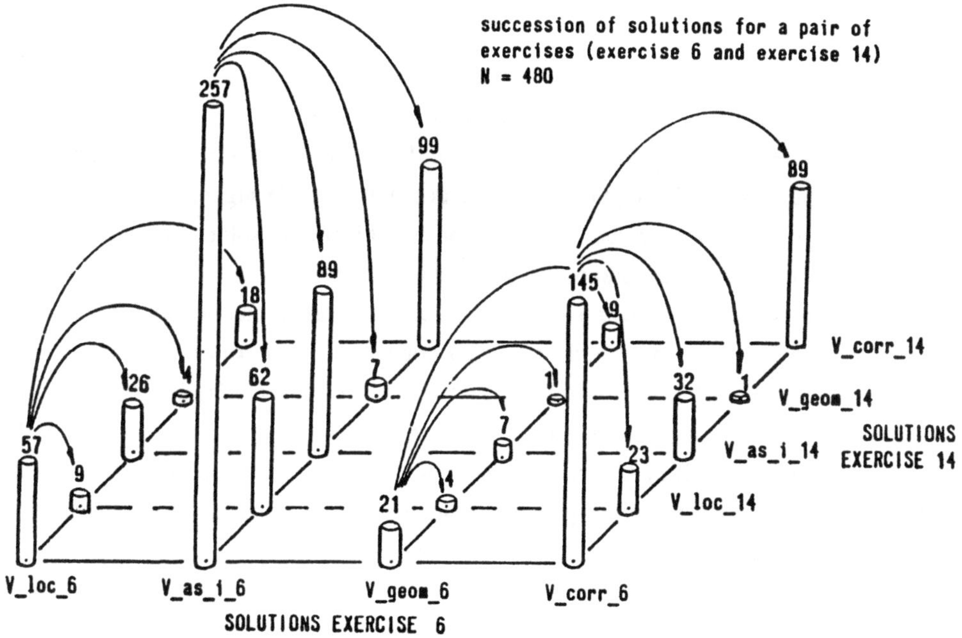

Figure 2: Distribution of the solutions in exercise 6 and redistribution of the solutions in the equivalent exercise 14, N=480. Solutions 3 and 4 in exercise 6 are collected in the same category V_as_I_6.

A correlation matrix of the 56 different students' conceptions and correct solutions to the test items contained only small values. An explanation for the missing correlations is suggested by looking at the distribution of solutions for an exercise of the first part of the test, and the redistribution of the solutions in the corresponding exercise in the second part (Fig. 2).
The distribution of solutions in exercise 6 shows a typical profile. The various sub-profiles obtained for the solutions to exercise 14 are very similar to each other in first approximation. This means that the selection of a solution in exercise 14 is nearly independent of the selection in exercise 6. This explains why the correlations between the different solutions show only small values.
It turns out, however, that a more detailed analysis of the results shows that there are significant patterns present, and that there is a relationship between these patterns and the students' ability to process information. To demonstrate this, the consistencies within individual pairs of questions (correct/correct or incorrect/incorrect) and the inconsistencies (incorrect/correct or correct/incorrect) were studied in more detail.

3. The Influence of Knowledge Structures on Information Processing

As a first step the students were grouped according to the number of correct solution pairs, and the number of students' conceptions used consistently in pairs of related exercises i.e., the equivalent conception had to be applied in both parts of the test. We suggest that the frequency of occurrence of pairs of correct solutions and of consistent incorrect solutions is an indicator of scientific and unscientific modes of knowledge structure.

In grouping the students as mentioned above, three "categories" can be defined (cf. Fig. 3): A small number (12 students) with six or seven correct pairs of solutions is classified as category 1, showing a "stable scientific understanding". The characterization "approaching a scientific understanding" is applied to category 2, which has 143 students. In this category there are more correct pairs of solutions than incorrect pairs. Category 3, which we characterize as "approaching an unscientific view of physics", comprises the rest of the students (325). The students in this category have the same number or more pairs of incorrect solutions than pairs of correct solutions. There was no student with more than five pairs of incorrect solutions. For this reason the category "stable unscientific view of physics" is missing.

The formation of the three categories is confirmed when the overall changes from incorrect to correct and from correct to incorrect solutions {Σ(ic-ci)} between the first and the second part of the tests were calculated for every group (cf. Fig.4).

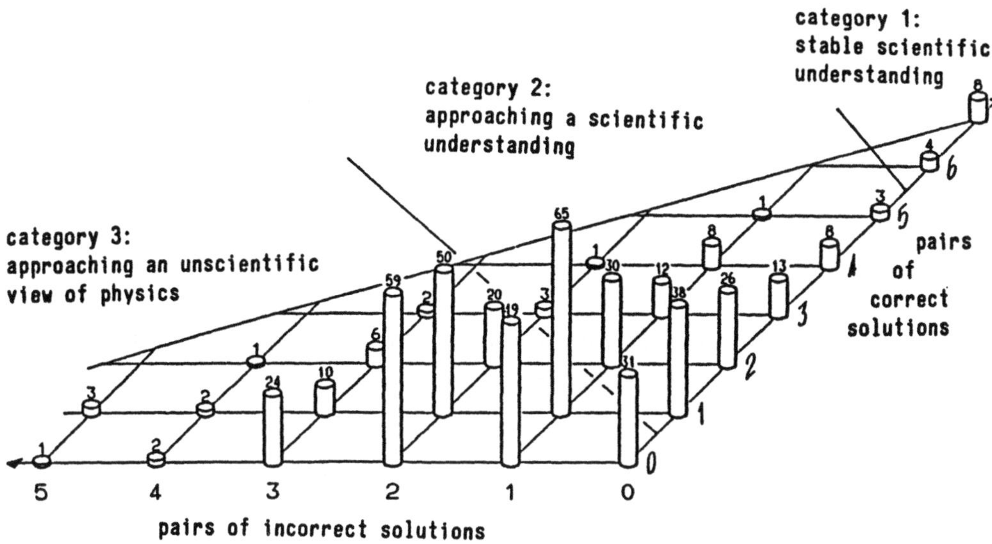

Figure 3: The variation of the number of students with pairs of correct solutions and pairs of incorrect solutions

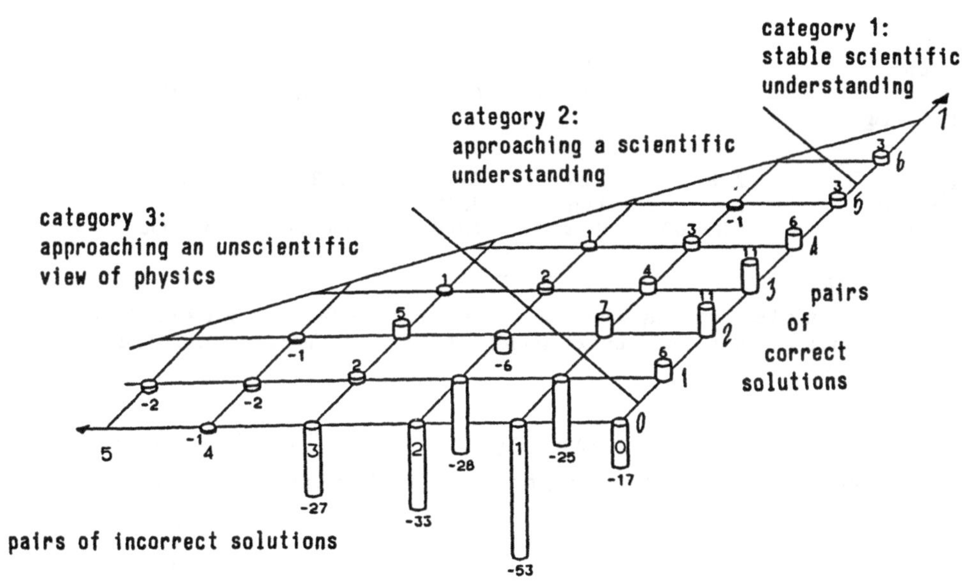

Figure 4: The term Σ(ic-ci) for groups of students with a certain number of incorrect and correct pairs of solutions

The changes $\Sigma(ic-ci)$ are positive for some groups and negative for others: in category 1 the term $\Sigma(ic-ci)$ is small, but positive. In this category the knowledge base of the students is already nearly perfect, and there is little opportunity for activation and learning processes to take place between part one and part two of the test. A stronger positive activation emerges in category 2. Here the results in part 2 are markedly better than the results in the first part, i.e., the students activate their knowledge of physics ("approaching a scientific understanding") and utilize the information given. In category 3 negative activations are predominant. The students in this category do not grasp the information given. They tend towards an unscientific view of physics and reactivate incorrect knowledge. The differences between category 2 and category 3 are statistically significant at the 1 percent level.

Version: / Category	no information	right/ wrong information	short information	detailed information	mean
category 1: stable scientific understanding	6.60 6.80 (5 stud.)	7.00 7.00 (3 stud.)	7.00 7.00 (2 stud.)	6.00 7.00 (2 stud.)	6.67 6.92 (12 stud.)
category 2: approaching a scientific understanding	3.15 3.62 (26 stud.)	3.65 4.00 (31 stud.)	3.37 3.70 (46 stud.)	3.37 3.75 (40 stud.)	3.39 3.76 (143 stud.)
category 3: approaching an unscientific view	2.18 1.50 (88 stud.)	2.25 1.66 (87 stud.)	2.40 1.99 (70 stud.)	2.32 1.75 (80 stud.)	2.28 1.71 (325 stud.)

Table 3: Mean number of correctly-solved exercises in the first part (first line: exercises 2 - 8) and in the second part (second line: exercises 10 - 16) of the test for the three categories and the four versions, N = 480

A more detailed analysis of the different categories is given in Tab. 3. Here the mean number of correctly solved exercises is listed for the first part (first line: exercises 2 - 8) and for the second part (second line: exercises 10 - 16) for every category, and for each feedback type. In category 1 and category 2 the results for the second part of the test are better in all versions than the results in the first part, i.e., activation and learning processes take place. In category 3 the reactivation of unscientific knowledge exceeds the learning process and leads to lower results in the second part of the test. For students in this category the knowledge base tends towards an unscientific view of physics, and it seems unlikely that these students will be able to process given information in a useful way. In all versions and categories the actual acquiring of knowledge remains far below the potentialities. It is clear that students' alternative conceptions exert a strong influence on the results on both parts of the test.

4. Discussion and Prospects

It is generally known that students' conceptions are context specific, i.e., they depend on the concrete task which is presented [1,2]. Often, in follow-up interviews, the students will comment on their solution and defend their conceptions. The interviewer may therefore get the impression that solutions and commentaries describe stable students' conceptions. The outcomes of our statistical analysis (cf. Fig. 2) lead to another interpretation: if the situation changes from a first exercise to a similar (from the physicist's point of view) exercise, the search for a solution to the second exercise starts from the beginning in first approximation. Instead of continuity we find change; instead of positive changes we find changes in both positive and negative directions. Since a psychological definition of learning includes positive change and stability of already correct conceptions, learning does not take place in this approximation.

Certain patterns reveal themselves in second approximation. The analysis of pairs of correct and incorrect solutions shows (cf. Fig. 3) that there are stable components. The distribution of students on the pairs of correct and incorrect solutions is seen to be nearly symmetrical. However, as we have defined it, the category "approaching an unscientific view of physics" comprises more than one half of the students. From this we would expect learning difficulties to arise, as is confirmed by the predominant negative changes in this category (cf. Fig. 4). We can interpret these findings as indicating that the influence of pre-existing cognitive structures is more important for the process of solution than the feedback information presented in the first part of our test.

The initiation of learning processes on a large scale was not achieved. Perhaps the feedback time was too short, or the expected transfer too large between the exercises of the first and the second part, or perhaps the time gap between the two parts of the test was too long. To clarify these questions additional investigations were carried out in a Technical Gymnasium with two feedback versions (right/wrong information, and short information). These investigations are still ongoing.

In a first variation the sequence of exercises was changed: exercise 9 followed the similar exercise 1 immediately, exercise 10 followed exercise 2, etc. The results were not very promising. In a second variation the students were given a review of the field of basic electricity before they sat the test. One version started with a recapitulation of the basic concepts, the second with four preliminary tasks. In both versions the students discussed their results with their class-mates and their teacher prior to doing the test. The four preliminary tasks challenged the students, and led to better results.

The best results with an increase in the solution frequencies of about 12% were achieved when the gap between the first and the second part of the test was closed by interposing additional exercises designed to reduce the required transfer and refresh the rules and concepts: Each of the eight exercises in the first part was followed by feedback, and then by a similar second exercise. Even in the interposed exercises, the results were already better. A more comprehensive introduction combining the eight interposed exercises and the four preliminary tasks produced nearly the same results, i. e., there are practically no supplementary effects produced by the four preliminary tasks. We interpret these results as a positive effect of the combination of feedback followed immediately by appropriate additional exercises. With the help of these components the student builds up the knowledge base. Additional interviews would clarify the extent to which the modified knowledge base is correct.

As a next step we are working towards improving student performance in the two still-difficult pairs of tasks, 4/12 and 8/16. This is again being done by interposing intermediate exercises, which we expect will lead to similar improvements in the frequency of correct solutions to tasks 12 and 16.

References

1. Delacôte, G., Tiberghien, A., Schwartz, J.: Recherche en didactique de la physique. Paris: editions du CNRS 1983
2. Driver, R., Guesne, E., Tiberghien, A.: Children's ideas in science. Milton Keynes: Open University Press 1985
3. Duit, R., Jung, W., and von Rhöneck, C. (Eds.): Aspects of Understanding Electricity. IPN-Arbeitsbericht 59. Kiel: Schmidt und Klaunig 1985
4. Shipstone, D.M., von Rhöneck, C., Jung, W., Kärrqvist, C.,Dupin, J.J., Joshua, S., Licht, P.: A study of students' understanding of electricity in five European countries. Int. Jour. of Sc. Educ. 3, 303-316 (1988)
5. VP-Expert. Paperback Software International 1987

Modelling the Learner: Lessons from the Study of Knowledge Reorganization in Astronomy

Stella Vosniadou

Aristotelian University of Thessaloniki, Greece, and University of Illinois at Urbana-Champaign, USA

Abstract: It is argued that conceptual change involves not only the addition of new information to existing conceptual structures but also the reorganization of existing conceptual structures and the reinterpretation of entrenched beliefs based on everyday experience. The more information we have about the knowledge acquisition process in a learner model the better chances we have of designing successful instructional sequences.

Keywords: learner component, conceptual change, restructuring

Introduction

The construction of intelligent learning environments requires rather diverse types of knowledge, ranging from general knowledge about computers, programming and artificial intelligence to specific knowledge about the particular domain in which the learning environment is to be constructed. In addition, what is needed is a model of the learner and of the learning process on which the instructional component must be based. In this paper, I will focus on the learner component of intelligent learning environments. More specifically, I will try to describe what I consider to be, from a psychological point of view, some of the essential characteristics of a learner model. My comments will be based on a series of studies of the knowledge acquisition process in the domain of astronomy which my colleagues and I have conducted cross-culturally (e.g., 2, 3, 5, 6, 7, 8).

In building a learner component, researchers usually rely on some general theory of learning, such as ACT theory, where learning is modeled in terms of production rules

NATO ASI Series, Vol. F 86
Intelligent Learning Environments
and Knowledge Acquisition in Physics
Edited by A. Tiberghien and H. Mandl

and changes in learning occur as a result of mechanisms such as generalization and discrimination. In this paper I will argue that although one can go very far with general theories of learning, these theories cannot always capture the peculiarities of the learning process as it occurs in specific knowledge domains and cannot explain the radical restructurings of knowledge which often characterize the knowledge acquisition process, particularly in the domains of physical science.

Many researchers attempt to solve the domain specificity problem by adding to the learner model information about the kinds of misconceptions a learner is likely to form in a given domain. There is no doubt that information about student misconceptions is very important in a learner component. However, this information is not sufficient by itself to lead to intelligent decisions about instructional strategies. Misconceptions usually represent attempts from the part of students to reconcile the scientific notions to which they are instructed with certain entrenched beliefs which are based on their everyday experience. In order to give up their misconceptions, students need to replace their entrenched beliefs with a different conceptual framework. It follows, that in order to be complete, a learner model must provide information about the entrenched beliefs that give rise to misconceptions in the first place. In addition, a learner component must contain information about the interrelationships among the various concepts that comprise a given domain because these interrelationships determine to a large extent the sequence of acquisition of these concepts.

2. Misconceptions and Entrenched Beliefs

It has become rather commonplace to talk about students' misconceptions in science, yet there does not appear to be a satisfactory attempt so far to explain how misconceptions are formed. The research my colleagues and I have conducted in the domain of observational astronomy has suggested that misconceptions represent students' attempts to reconcile the scientific explanation of a phenomenon, to which they are exposed through instruction, with certain entrenched beliefs which are based on everyday experience. For example, our studies of elementary school children's ideas about the shape of the earth (5, 7), have shown that the great majority of elementary school students use in a consistent fashion the small number of mental models of the earth shown in Table 1.

As can be seen in Table 1, some elementary school students think that the earth is flat like a rectangle or flat like a disc; others think that there are two earths: a flat one on which people live and a spherical one which is up in the sky. Some children believe that the earth is a hollow sphere with people living on flat ground inside it, and so on.[1] These mental models can be grouped into three categories: scientific, initial, and synthetic.

Scientific mental models are the models which are consistent with the currently accepted scientific views. The mental model of a spherical earth is a scientific mental model. *Initial* mental models show no influence from the scientific views. They appear to be based on students' everyday experience as this experience is interpreted by the human cognitive apparatus. The mental model of a rectangularly shaped flat earth can be considered an initial mental model. *Synthetic* models are those that reveal some combination of scientific and initial models. This kind of mental model comes closer to what other researchers have called misconceptions. They reveal some misinterpretation of the scientific ideas to which students are exposed through instruction. Synthetic mental models are those of the dual earth, the hollow sphere, and the flattened sphere.

In earlier work I have shown that all the synthetic models of the earth can be interpreted as attempts to reconcile the information coming from adults that the earth is a sphere with two entrenched beliefs: (a) the belief that everything that appears to be flat is indeed flat (and therefore that the ground is flat), and (b) the belief that unsupported things will fall "down" (and therefore that both the earth and the people and objects on the earth need to be supported).

In forming synthetic models, children change their initial models in a way that allows them to retain some or all of their entrenched beliefs, without contradicting adult teachings. For example, the children who change their initial model of a rectangular earth to the synthetic model of a dual earth have retained almost all of the entrenched beliefs that gave rise to the initial model in the first place. These students answer our questions in such a way that shows that they still believe that the ground is flat, that the earth is supported by something (ground or water), and that the sky is located only

[1]The methodology used to derive students' mental models is described in detail in Vosniadou & Brewer, 1989. Briefly, this methodology consisted of asking students as many questions as possible investigating a given concept, and then checking to see whether their individual responses to these questions could be explained by assuming that they used one of these mental models in a consistent fashion.

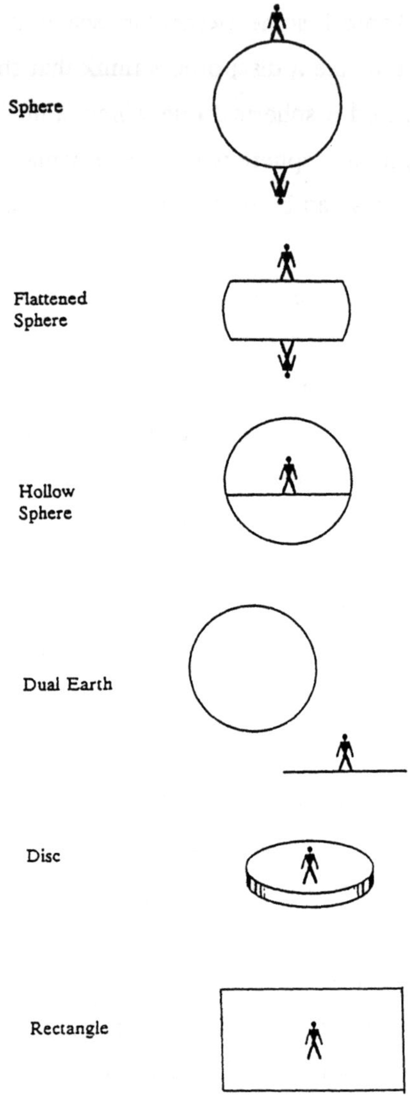

Table 1. Mental Models of the Earth

above the earth. The only thing that has changed is that they have added to their knowledge base the information that there is a planet called earth which is round like a ball and which is located up in the sky.

A detailed examination of children's responses reveals that there is a progression of more and more synthetic models depending on how many of their entrenched beliefs children have given up. For example, the hollow sphere synthetic model is more advanced than the dual earth model. Children who hold this model have given up their belief that the earth needs to be supported. These children conceptualize the earth as a sphere suspended in space. They still believe, however, that the ground is flat and that the people and objects on the earth need to be supported. For this reason they create the misconception that people live on flat ground inside the hollow sphere.

If I am correct in my analysis that misconceptions are formed because students try to reconcile scientific notions with their entrenched beliefs, it is important, from an instructional point of view, to identify these beliefs and to help children replace them with different, more adequate explanations. Instruction which focuses *only* on the misconceptions without consideration of the entrenched beliefs that underlie them is not likely to be successful. Consider for example the case of the child who believes that the earth is a hollow sphere. Telling this child that the earth is not a hollow sphere and that people live on all sides of the earth on the outside, many disrupt his or her hollow sphere model but will not, most probably, produce a correct scientific model. Children create the hollow sphere model because they do not understand how it is possible for people to live at the sides and bottom of the earth without falling and how it is possible for the ground to be flat when it appears to be round. For instruction to be successful we must address not only the misconceptions that students have but also the entrenched beliefs that lurk behind these misconceptions and give rise to them in the first place.

Identifying and addressing students' entrenched beliefs is also a more efficient and parsimonious way of dealing with misconceptions than addressing each misconception individually. This is the case because a very small number of entrenched beliefs can lie at the root of a large number of misconceptions.

It follows from the above discussion that it is important when we construct the learner model of an intelligent learning environment to include information not only about students' misconceptions but also about the entrenched beliefs that cause these misconceptions. I believe that the success of designing learning environments which promote conceptual restructuring depends to a large extent on the success of (a) our basic research to identify the entrenched beliefs that cause misconceptions, and (b) our instructional methods to make children reinterpret these beliefs.

3. Interdependencies Among the Concepts that Comprise a Given Domain

The research my colleagues and I have conducted has shown an interdependency among the various concepts that comprise the domain of astronomy. For example, students' mental models of the earth appear to constrain their mental models of the sun and the moon, and their explanations of the day/night cycle. Students who lack the concept of a spherical earth cannot understand correctly the scientific explanation of the day/night cycle in terms of the earth's rotation. Table 2 presents two misconceptions of the day/night cycle which could be contributed to a synthetic model of the earth. The child whose model is shown in Table 2A had constructed the earth as a disc. Although this child understood that the disc earth turns in circles, he (justifiably) could not see how this movement explained the disappearance of the sun at night. In order for his model to have any explanatory power he had to also assume that the sun moves down and hides behind the mountains.

(a) (b)

Table 2. Explanations of the day/night cycle in terms of the earth's axis rotation for (a) a disc earth mental model and (b) a hollow earth mental model.

Another child with a hollow sphere model created the misconception of the day/night cycle shown in Table 2B. According to this model, the hollow-sphere consists of two hemispheres; the lower hemisphere which represents the earth on which people live, and the upper hemisphere which represents the sky covering the earth like a dome. The sun and moon are located at two opposite sides--a day side and a night side--within the upper hemisphere. As the earth turns people move from the night side to the day side!

As these examples demonstrate, students interpret the scientific explanation of the day/night cycle in the context of the mental models of the earth they already have. When these mental models differ from those required for the accurate interpretation of the new information then the new information is very likely to be misrepresented to fit the existing model. One can imagine, that as a domain becomes more and more complex, the possibilities for misunderstandings multiply and conceptual change requires the reorganization of larger and larger parts of underlying structure constructed on the basis of narrowly-built entrenched beliefs. For this reason it is important when building a learner model to know the sequence in which the concepts that conprise a given domain are acquired and to try to design environments where the knowledge acquisition process is consistent with this sequence.

4. The Enhancement of Conceptual Restructuring

General theories of learning account for conceptual change via mechanisms such as generalization, differentiation, coalescence, and hierarchical integration which operate on existing conceptual structures. These types of mechanisms may be adequate to account for conceptual change in many cases but not in all. As I have argued, the knowledge acquisition process in the domain of astronomy requires the reformulation of existing mental models and the reinterpretation of entrenched beliefs. When our models are revised or our beliefs are reinterpreted the observations which gave rise to them in the first place do not change. In our everyday experience, the ground still appears to be flat and things still fall in a downward direction when not supported. What changes is the interpretation of these observations. This change in interpretation is an important characteristic of theory change (e.g., 1).

Little is known so far about the exact mechanisms that promote this type of conceptual restructuring. Is it better for students to be told the new explanation explicitly or is it better to be put in a situation where it is possible to discover it? What is the importance of the social and cultural context in which new learning is acquired and how much the creation of a cognitive conflict situation promotes conceptual change? These are important instructional questions and more research is needed before answering them. At this point I can only speculate about some characteristics of learning environments which may enhance conceptual change.

1. Providing clear explanations of scientific concepts. One such characteristic would appear to be the availability of clear explanations of the scientific concepts, preferably in the form of conceptual models or analogies. One would think that this is the minimum requirement one can place on instruction and yet it is a requirement that is hardly met. An in-depth analysis of the astronomy units in four widely used science series for elementary school students I have recently conducted showed that crisp and clear explanations are rare (e.g., 4). While it is certainly the case that better explanations of scientific concepts can be provided in science texts, or in class instruction, computer-based learning environments have a distinct advantage over traditional methods in their capability to provide clear explanations of scientific concepts. This is because through computers we can model in a dynamic way highly abstract processes and phenomena. They can show us how the planets move in the solar system or how the blood circulates in the body. They can take us to the moon and show us how the earth looks from there, and so on. They have tremendous possibilities to make the unobservable observable.

2. Extending students' experiences. Again, this can be done with traditional methods. Texts can be written with pictures that show the moon in the sky during the day, or the teacher can take the children who believe that the moon causes night and show them that the moon can be in the sky during the day. However, computer-based learning environments have a distinct advantage over traditional methods. Through microworlds and simulations children can have experiences that are very difficult if not impossible to have in the real world, such as the experience of operating in a frictionless environment. Unfortunately, it is often the case that simulations are presented as isolated instances distinct from what students experience in the everyday world. These kinds of presentations do not allow students to understand how the simulated environments relate to their existing observations and thus may not promote conceptual reorganization.

3. Creating metaconceptual awareness. In my view an important obstacle to conceptual change is children's lack of metaconceptual awareness. While children construct explanations of phenomena they do not seem to be aware of the theoretical nature of their activity. They seem to believe that these explanations represent the real state of affairs about the world. This can be inferred from the fact that when children are presented with scientific ideas which are contrary to their beliefs they do not question these beliefs. Rather, they try to assimilate the scientific notions to their existing beliefs by creating synthetic models or misconceptions. It is important when we teach science to provide children with situations which make them realize that their beliefs about the world are not true facts but theoretical constructions which may be subject to falsification.

This could be done by creating learning environments which present students with empirical observations which are not consistent with their entrenched beliefs and showing them that if these beliefs are modified their empirical adequacy will increase. Learning environments which aim at enriching students' experiential knowledge without making them aware of the constructive nature of their knowledge building activities fail to create this necessary metaconceptual awareness.

5. Conclusions

I have argued that when we design intelligent learning environments in the domain of physics we must pay particular attention to the representation of students' entrenched beliefs. Misconceptions are caused because students try to reconcile scientific notions with certain beliefs they have constructed on the basis of their everyday experience. In order for conceptual reorganization to be achieved these entrenched beliefs need to be reinterpreted.

The reinterpretation of entrenched beliefs is not achieved with mechanisms such as generalization and discrimination which general theories of learning usually adopt. It requires that students be provided with new observations that make it possible for them to enrich their knowledge base and understand the limitations of their existing beliefs. Intelligent learning environments have certain advantages over traditional methods for the promotion of belief reinterpretation.

Finally, I have argued that the concepts that comprise a given domain are interrelated and that there is a sequence in their acquisition. There is reason to believe that learning environments which present concepts in a sequence consistent with that of their acquisition will be more successful in producing conceptual change than the ones that are not.

Author Notes

The research reported in this paper was supported in part by a grant from the National Science Foundation, BNS-85-10254, and in part by a grant from the Office of Educational Research and Improvement under Cooperative Agreement No. G0087-CI001-90 with the Reading Research and Education Center. This publication does not necessarily reflect the views of the agencies supporting the research.

1. Kuhn, T.S. (1962). *The Copernican Revolution*. Cambridge, MA: Harvard University Press.

2. Vosniadou, S. (1987, April). *Children's acquisition and restructuring of science knowledge*. Paper presented in a symposium on "Children's procedural knowledge in science" at the annual meeting of the American Educational Research Association, Washington, DC.

3. Vosniadou, S. (1989). One the nature of children's naive knowledge. *Proceedings of the 11th Annual Conference of the Cognitive Science Society*, Ann Arbor, Michigan.

4. Vosniadou, S. (1991). Designing curricula for conceptual restructuring: Lessons from the study of knowledge acquisition in astronomy. *Journal of Curriculum Studies*.

5. Vosniadou, S., Archodidou, A., & Kalogiannidou, A. (in preparation). *Mental models of the earth in children from Greece: A cross-cultural study*.

6. Vosniadou, S., & Brewer, W.F. (1987). Theories of knowledge restructuring in development. *Review of Educational Research*, *57*(1), 51-67.

7. Vosniadou, S., & Brewer, W. F. (1989). *The concept of the earth's shape: A study of conceptual change in childhood* (Tech. Rep. No. 467). Urbana-Champaign: University of Illinois, Center for the Study of Reading.

8. Vosniadou, S., & Brewer, W.F. (1990). A cross-cultural investigation of children's conceptions about the earth, the sun, and the moon: Greek and American data. In H. Mandl, E. DeCorte, N. Bennett, & H.F. Friedrich (Eds.), *Learning and instruction: European research in an international context* (Vol. 22, pp. 605-629). Oxford: Pergamon.

Cognitive Theories as a Basis for Student Modelling

Heinz Mandl[1] and Aemilian Hron[2]

[1]Universität München, Institut für Empirische Pädagogik und Pädagogische Psychologie, Leopoldstr. 13, D-8000 München 40, FRG

[2]Deutsches Institut für Fernstudien an der Universität Tübingen, Arbeitsbereich Lernforschung, Konrad-Adenauer Allee 40-44, D-7400 Tübingen 1, FRG

Abstract: Starting out from Anderson's concept of procedural learning and White and Frederiksen's approach of mental models, student modelling in the domain of intelligent tutoring systems is discussed with regard to cognitive approaches. Anderson's approach provides a conceptual basis for analyzing knowledge acquisition processes with regard to cognitive skills. Within the framework of this approach Anderson models problem-solving behaviour and knowledge acquisition processes based on the description of an ideal student behaviour on the one hand (expert model), and incorrect behaviour on the other hand (bug library), applying production rules. White and Frederiksen focus primarily on conceptual knowledge and relate to the construct of mental models. Knowledge acquisition processes are modeled as a progression of mental models which the student passes through in the course of acquiring expertise. A more critical perspective reveals that Anderson lays emphasis on procedural learning, completely neglecting cognitive processes. The weak points of White and Frederiksen's approach consist in the fact that knowledge acquisition processes are not described in psychological terms, but rather in physical terms. Although the above-mentioned ways of approaching the matter of knowledge acquisition differ widely, cognitive approaches allow a more differentiated analysis of the representation and acquisition of knowledge, thus providing an important basis for student modelling.

Keywords: cognitive theories, student modelling

NATO ASI Series, Vol. F 86
Intelligent Learning Environments
and Knowledge Acquisition in Physics
Edited by A. Tiberghien and H. Mandl

1. Introduction

Student modelling is a central characteristic of Intelligent Tutoring Systems (ITS). ITS is a new form of computer-based learning program which aims at adapting the computer-based instructions to the prior knowledge and learning progress of the individual student. For this purpose, during the course of instruction, the program constructs a model of the individual understanding that the student has from the learning material. The fundamentals of student modelling are the way in which a student approaches the answering of questions, the solutions of problems, and the call for help or other input within the program. On the basis of that approach, a data structure, which is named student model, is then developed. A student model represents the student's current knowledge to which the instruction is to be adapted.

The characteristics of student modelling reveal the fundamental embedding of this issue in research of cognitive psychology. ITS is a relatively new research area which, from the beginning, (together with the Artificial Intelligence methods), has been run on the lines of cognitive approaches. ITS originates from the pioneering work of Carbonell [12], Collins & Quillian [13] and others. In the beginning, research was mainly concerned with problems of knowledge representation, as can be clearly seen in the publication of Sleeman and Brown [31]. Examples are the works of Burton [11] concerning procedural networks, Goldstein's [17] genetic graphs, or the qualitative simulations in SOPHIE [10]. Student modelling, in the sense of cognitive modelling, has since come very much to the fore. In this context, research interest has focused on knowledge representation which corresponds with the student's mental procedures in the respective knowledge domains (see e.g. [21, 24, 25]).

Basically, student modelling can be regarded as a psychological issue, because in any case assumptions about the student's knowledge are made. However, in the framework of the development of an ITS, the psychological issue is closely linked with the issue of the technical effort required for the desired degree of fine-grained modelling. There are various student modelling techniques (see [32]) which pose different psychological research questions.

The most widespread technique is the so-called "overlay-approach" in which the student's knowledge structure is described in terms of a representation of the subject matter in an expert model. The student's respective knowledge is considered to be a subset of a domain expert's knowledge. Learning is regarded as a process of increasing acquisition of expert knowledge, without taking the preliminary stages of expert knowledge, misconceptions, or distorted knowledge into consideration. Overlay models require the structuring of a knowledge domain and its computational representation. This type of model is very common in the domain of ITS [17].

As opposed to overlay models, error description models are explicitly directed at the student's erroneous procedures and misconceptions. Modelling ensues on the basis of an error library, that is, a list of possible errors: with any particular student it is possible to compute which error (or combination of errors) accounts best for his or her incorrect response. A detailed form of that approach is the reconstruction of bugs. Bugs can be explained in terms of prespecified primitives of buggy procedures. The very primitives are no bugs, but neutral with respect to correctness. The basic assumption is that errors observed in student's performance can be accounted for by tracing the errors back to a combination of primitives. The set of (prespecified) primitives is critical for systems based on this paradigm [26]. Approaches based on reconstruction of bugs have been used mainly for research purposes (e.g. [30, 20, 22]). The most sophisticated forms of student modelling are runable simulations of learning behaviour, so-called "process models". In that case, it is necessary to describe the learner's mental representation and operations. For this purpose, detailed models of cognitive structures and processes are afforded which can be assumed to underlie the relevant skills or knowledge. This student modelling does not only tend in the direction of showing which correct or incorrect concepts are applied by the student, but also how he or she combines concepts to a more or less correct solution which is in mind, and how plans are performed. Of course the justification for the effort in the course of the development of such a student model component of an ITS is questionable: a fine-grained student model requires probably more work than it is worth, if the tutoring cannot be adapted to the intricacies of a particular student's misconceptions [32].

In the following, two instances of student modelling will be presented which are oriented towards the latter form of student modelling. Questions of technical effort and suitability will not be discussed; relations between student modelling and cognitive theories will, however, be worked upon. On the one hand, the student modelling developed in the framework of the LISP-tutor by Anderson and coworkers, which refers back to the the ACT*-Theory [2], will be presented. On the other hand, the student modelling of White and Frederiksen [34, 35] in the context of the intelligent learning environment QUEST, which is based on the approach of mental models and qualitative reasoning, will be discussed.

2. The architecture of cognition: Anderson's concept

2.1 Theoretical approach

Anderson's student modelling, which underlies the Intelligent Tutoring Systems developed by him and coworkers, is based upon the ACT* theory [2]. ACT* stands for "adaptive control of thought" and presents the consequent continuation of many previous versions (e.g. [1]).

Anderson, with his theory, has devised an architecture of cognition which should reflect a multitude of basic phenomena of human information processing. The theory is concerned with higher-order cognitive processes such as memory, speech, problem-solving, imagery, deduction, and induction. The implementation of the postulated cognitive system is widely based upon the application of production systems which came from computer science and have been adopted by cognitive psychology.

The principal constituents of Anderson's cognitive system are working memory, declarative and procedural memory. The working memory contains the directly accessible (activated) information which stems either from the declarative memory or is encoded from the environment. Declarative and procedural memory differ not only through contents (namely factual knowledge versus active knowledge) but also through the organisation of that memory contents. The declarative memory, as form of representation, comprises a propositional network, temporal strings, and spatial concepts. In the procedural memory, the possible system's actions are stored in form of productions. Productions are condition-action units. Actions can be carried out, if data patterns contained in the working memory correspond with productions contained in the production memory.

In ACT*, Anderson assumes that knowledge is originally present in declarative form and that it is gradually summoned and applied by means of general procedures. However, if a certain activity is repeated to a certain degree, then - through automatic actions (proceduralisations) - this type of use is gradually replaced by other, newly created productions. The latter can carry out that activity directly. If several of these automatic actions are condensed - as a series of productions - to a single and thus complete production, this is called a combination. Proceduralisation and combination are called knowledge compilation and make it possible to explain how a faster and safer execution of activities can be attained. New productions can be gained from generalisation or discrimination of conditions.

2.2 Student modelling

ACT* forms the theoretical basis for the Intelligent Tutoring Systems developed by Anderson and coworkers. As an example, the LISP-Tutor [5] will be referred to here. This ITS permits the student to acquire the programming language LISP in a similar way to a human coach providing individual instruction.

It gives instructions on the screen, assigns tasks, and continually follows the task handling of the student. As long as the student does not make an error and progresses with learning, it stays in the background. If problems occur, it intervenes and gives advice and help.

The student can also demand information and help from it at any time. Otherwise, the program proceeds according to the principle that the student should achieve as much as possible without any external help.

The learning environment is linked with a student model which is based on a detailed analysis of each portion of the student's solution in the specific domain of LISP. This model is able to diagnose a correct or incorrect approach to solution, in order to control the tutorial dialogue [27, 6]. In order to assess the student's progress, the tutor includes the domain knowledge necessary to solve problems (ideal student model) as well as knowledge about common mistakes and poor strategies of novices (bug catalogue).

The domain knowledge represents the programming knowledge an ideal student uses in solving LISP problems. This ideal model is based on ACT* theory applied to the analysis of learning how to program [7]. The ideal student model is implemented in GRAPES, that is, Goal-Restricted Production System [29]. The architecture of GRAPES uses goal decomposition, found in solving programming problems. Each production in the ideal student model includes a goal specification that should reach the rule and a description of conditions in which the rule is applicable. The ideal student model includes planning and coding production rules. The planning rules outline an algorithm in order to attain a specific program specification. The coding productions then write the code in order to produce the algorithm.

There are a lot of buggy rules (buggy catalogue) associated with the production rules in the ideal model and these present misconceptions which novices in programming often develop whilst learning. Buggy rules are incorrect variations of the correct rules in the ideal model [9, 30]. Buggy rules are poor strategies, semantic confusion of the basic LISP functions, misunderstandings about manipulation objects, such as variables, erroneous use of syntactic constructs (e.g., missing quotes, misgrouped parentheses), or other common slips [27].

In order to diagnose errors, the tutor compares the student's input with the ideal model's rules and the associated buggy rules which are responsible for the actual state of problem solving. The combined use of the ideal and the buggy model (i.e., the generic model) is what defines the so-called model-tracing methodology: the tutor traces out the path the student endeavours to take through the generic model and insists that the student stays on the correct path.

Based on empirical evidence that the best moment to correct a student's misconception is the moment of its occurence, the LISP-tutor responds to any minimal distortion the student is intending. Thus the tutor is able to construct a more accurate model of the student's reasoning. The diagnostic procedure is based on the fact that the student receives immediate feedback to any error; thus, further incorrect procedures are excluded.

Reiser, Anderson and Farrell [27] have shown that the tutor contains 375 production rules in the ideal student model and 475 buggy versions of these rules. These rules enable the tutor to diagnose and respond appropriately to about 45 to 80 percent of the student's errors, which depends on the complexity of the lesson. Well-worked lessons diagnose 80 percent of the student's errors, while newly implemented lessons only diagnose 45 percent of the errors.

In a study from 1990, Anderson [4] emphasized that the analysis of student behaviour is consistent with the ACT* production system analysis of skill acquisition. During the initial phase of the study a remarkable increase in performance could be noticed which diminished in the sequel. This observation is known as the "power law of skill acquisition". Anderson attributes the initially heavy increase in performance to the compilation of knowledge taking place, the subsequently reduced increase to tuning.

3. Mental models and qualitative thinking

3.1 Theoretical approach

White and Frederiksen's concept of student modelling - developed in the framework of the intelligent learning environment QUEST - is based upon the theory of mental models and qualitative reasoning. Mental models describe a theoretical construct which relates to the internal representation of structure and function of complex systems. Mental models guide the understanding and handling of a system. They enable to explain how a system generates system-states and to derive prognoses about the system behaviour and the effects of input into the system [19].

The approach of mental models presents a relatively new research topic for cognitive psychology: which knowledge does an individual hold in mind about a specific topic and how is this knowledge used in order to make inferences or solve problems, and so forth. Mental models research shifts the emphasis onto specific content domains and postpones general architectural questions about the representation of knowledge in memory, or any search for content-free rules of thought [23]. Thus, the focus lies less on procedural domains, such as mathematical skills or algebraic manipulations, but rather on richly semantic domains, such as physics, geography or medicine. Knowledge of such domains not only includes procedural skills, but also the understanding of concepts and principles and their usage. Additionally, the student's misconceptions, which are frequently developed in daily life, are equally relevant.

The psychological approach of mental models can be traced back to Craik [14] and Johnson-Laird's [18] fundamental further development of this theory. In principle, mental models are distinguished from other forms of representation in so far as they share the relational structure of the domain they represent. Mental models are concrete, analogical representations contrary to propositional representations which, for example, can be found in the framework of semantic networks [28]. In Artificial Intelligence research, this approach has been further developed in a sophisticated way with the aim of simulating human expertise. The focus is on the representation of qualitative models which are presented on the basis of various theories and qualitative vocabularies [8].

One major work of this kind, upon which White and Frederiksen's student modelling is orientated, is the theory of mechanistic mental models developed by de Kleer and Brown [15]. The core of this theory is the assumption that the physical world is considered as a complex mechanism. Accordingly, physical thinking should be describable in terms of a theoretical description of the structure and function of each of the components of a physical system or device. A problem-solver reasons qualitatively by means of analysing each of the cause-effect connections between the device-components and by developing a model of its causal functioning from which he or she can explain or predict the device's total behaviour.

White and Frederiksen apply this theoretical approach to the domain of electrical circuits. They represent, on this basis, the different steps of the emergence of expertise by means of a progression of increasingly sophisticated models. Furthermore, White and Frederiksen's approach includes an interesting aspect concerning the development of a computer-assisted learning environment: a model does not only deliver a hint about the student's knowledge acquisition, but simultaneously guides the interactive simulation of the relevant domain phenomenon.

3.2 Student modelling

The intelligent learning environment QUEST (Qualitative Understanding of Electrical Trouble-Shooting) is an offspring of the Intelligent Tutoring Systems SOPHIE [10] and STEAMER [16]. It allows interactive simulations and qualitative explanations and possesses an expert system for trouble shooting. Basically, QUEST is based on a series of increasingly complex qualitative models which correspond with the desired development of the student's mental model. This model progression is embodied in curriculum form. The learner can successively work through each problem according to the progression and thus acquire knowledge in a structured manner. He or she also has the possibility of free choice or can set up specific circuits or problems. A problem solving environment is put at the student's disposal wherein he or she can call up completed circuits, modified circuits or build and test them.

The system is able to generate functional, qualitative causal models for the circuits which the learner creates. On this basis, the learner can demonstrate the behaviour of a circuit and have it explained - both through written notice of the relevant rule and through corresponding verbal output. While debugging errors, the learner can also find help by an expert system.

The central characteristic of student modelling consists in the assumption that at any level of learning the qualitative model which underlies the problem and the computer simulation represents the mental model acquired by the student. Thus, learning is regarded as a process of model transformation: whilst acquiring an expert model, the student forms a series of models from which one is always suitable for a sub-category. As a reaction to the demands of more complex problems, which the student is tackling, these models are transferred to increasingly more suitable models. The model transformation can include addition, modification, differentiation or generalisation of model components, or even the construction of alternative models [33, 34].

In principle White and Frederiksen [33, 34] differentiate between three dimensions in which models can vary: perspective, order and degree of elaboration.

A model's perspective relates to the way of thinking with regard to the description of the circuit's behaviour. One can consider:

1) The superordinate function of a circuit; for example, its purpose and how subsystems coordinate within a circuit in order to reach a goal, that is, *functional models.*

2) The behaviour of a circuit on the level of circuit components, that is, how changes in the state of one of the devices can cause modifications in the other devices, that is, *behaviour-related models.*

3) The behaviour of a circuit at microlevel, that is, how electrical currents are spread over a resistance when the voltage drop changes in relation to the resistance, that is, *reductionist, physical models.*

Both models are qualitative as well as causal, but each represents a circuit from a different point of view. The spreading of effects of modifications in the circuits ensues by means of applying different laws accordingly (i.e., functional interrelations versus Kirchhoff's Law of Voltage versus Law of Coulomb). The models are thus directed at modelling various aspects of the domain phenomenon.

The dimension of order relates to *behaviour-related models*. White and Frederiksen distinguish between models of zero order and of *first order*. Zero order models are able to think about binary states of the device and can answer questions such as: "In this connection, is there a voltage drop in the light source and is, as a consequence, the light on or off?" Whereas first order models think on the basis of the qualitative (first order) derivations and can answer questions such as: "Is there an increase in the voltage of a component when the resistance of another component is reduced?"

Whilst learning, qualitative models can increase with regard to their degree of elaboration. This is controlled by the number of qualitative rules which are used to spread the effects of changes in the conditions of the connection components.

In sum, it is obvious that White and Frederiksen's student modelling rests on a sophisticated description of a progression of qualitative models of the contents to be imparted. Furthermore, these qualitative models monitor the interactive simulation of electrical circuits and provide qualitative explanations. In this sense, student modelling is closely tied to the presentation of content, since target mental models serve as a basis of knowledge representation.

4. Final evaluation

One cannot intend sophisticated student modelling without taking reference to cognitive approaches. Both instances described before reveal how that problem could be tackled in different ways.

Anderson's [2] approach is based on a cognitive theory which provides a generally valid statement on the development of cognitive structures and processes. That theory offers mechanisms for student modelling concerning the analysis and description of problem solving and knowledge acquisition. According to the claims of this theory, these processes are not domain-specific. Basically, this theory is orientated towards skill acquisition. It meets several of the most important criteria for student modeling: it allows the expression of skills at various levels of expertise within the same representational framework and supplies dynamics which allow novice representations to gradually grow into expert representations.

In Anderson's student modelling, these features are only applied conceptually in connection with the "analysis of the domain of learning how to program" and the respective development of conceptual models. The modelling is based on prespecified production rules which are not refined; Anderson's learning algorithms are not implemented.

According to Payne [23], Anderson's theory shows three fundamental problems: it neglects context effects, it neglects to model mistaken performance, and it neglects the role played by rich conceptual representations. Even skills with a large procedural component rely on a great deal of conceptual knowledge, so that the content and structure of the conceptual knowledge may continue to develop with expertise, alongside procedural knowledge.

On the background of these problems, Anderson has continued to develop the ACT* Theory. The new version was named PUPS (Pen Ultimate Production System) [3]. That theory is especially directed at a detailed explanation of knowledge acquisition from the point of view of intake of declarative knowledge into working memory. Thus the PUPS theory is mainly concerned with the acquisition of new declarative knowledge which ensues by means of processing experience through existing productions (e.g., language comprehension) and application of declarative knowledge within novel situations (i.e., situations for which productions do not exist) by means of analogy and pure search. That approach, which considers processes of understanding in connection with the encoding of declarative knowledge, could certainly benefit student modelling. However, being novel, these considerations have not been taken into Anderson's student modelling.

According to White and Frederiksen [33, 34], student modelling is based on the approach of mental models and qualitative thinking in the knowledge domain of physics. The theoretical considerations are not orientated towards general explanations concerning psychological mechanisms, but considered to be the result of direct acquisition and modification of physical knowledge, in this case, of knowledge about electric circuits.

The approach aims at understanding the interplay of qualitative reasoning, human expertise and instruction, especially in the domain of physics. It is based on conceptual knowledge. Knowledge acquisition is described as a hierarchy of mental models including its related procedures. Surprisingly, White and Frederiksen do not use bug models, which, in the context of the notion of misconception, is normally intended in semantically rich domains. However, White and Frederiksen claim to use a detailed organisation of the knowledge domain and of knowledge modification. Thus, a bug model does not seem necessary for them. Concerning general psychological epistemic interest, that approach certainly seems to be restricted, because cognitive processes, for example, processes of automation of knowledge, chunking or memory capacity, are ignored.

The discussion of these different approaches points out, that cognitive approaches must be taken into account to structure the domain and to represent processes of knowledge acquisition. However, there are different kinds of approaches that are focussing on specific aspects depending on theoretical assumptions and characteristics of knowledge domain.

Moreover, there is the problem of translating the postulated mechanisms and mental models into running simulations which are sensible in the framework of on-line diagnosis. However, cognitive approaches provide a good base for student modelling.

References

1. Anderson, J.R. : Language, memory and thought. Hillsdale, N.J.: Erlbaum 1976.
2. Anderson, J.R. : The architecture of cognition. Cambridge, M.A: Harvard University Press 1983.
3. Anderson, J.R. : A theory of the origins of knowledge. Artificial Intelligence, 40, 313-351, 1989.
4. Anderson, J.R. : Analysis of Student Performance with the LISP Tutor. In N. Frederiksen, R. Glaser, A. Lesgold & M.G. Shafto (Eds.), Diagnostic Monitoring of Skill and Knowledge Acquisition (pp. 27-50). Hillsdale, N. J.: Erlbaum, 1990.
5. Anderson, J.R. & Reiser, B.J. : The LISP tutor. Byte, 10 (4), 159-175, 1985.
6. Anderson, J.R., Boyle, C. & Yost, G. : The geometry tutor. Proceedings of Ninth International Joint Conference on Artificial Intelligence (pp. 1-7). Los Altos, CA: Morgan Kaufmann, 1985.
7. Anderson, J.R., Farrel, R. & Sauers, R. : Learning to program in LISP. Cognitive Science, 8, 87-129, 1984.
8. Bobrow, D.G. (Ed.) : Artificial Intelligence. Special volume on qualitative reasoning about physical systems. Amsterdam: North-Holland, 1984.
9. Brown, J.S. & Burton, R.R. : Diagnostic models for procedural bugs in mathematical skills. Cognitive Science, 2, 155-191, 1978.
10.Brown, J.S., Burton, R.R. & de Kleer, J. : Pedagogical, natural language, and knowledge engineering techniques in SOPHIE I, II and III. In D. Sleeman & J.S. Brown (Eds.), Intelligent Tutoring Systems (pp. 227-282). New York: Academic Press, 1982.
11. Burton, R. : Diagnosing bugs in a simple procedural skill. In D. Sleeman & J.S. Brown (Eds.), Intelligent Tutoring Systems (pp. 157-184). New York: Academic Press, 1982.
12. Carbonell, J.G. : AI in CAI: an artificial intelligence approach to computer-assisted instruction. IEEE Trans. Man-Mach. Syst., 11, 1970.
13.Collins, A. & Quillian, M.R. : Retrieval time from semantic memory. Journal of Verbal Learning and Verbal Behavior, 6, 240-247, 1969.
14.Craik, K. : The nature of explanation. Cambridge: Cambridge University Press, 1943.
15. De Kleer, J. & Brown, J.S. : Assumptions and ambiguities in mechanistic mental models. In D. Gentner & A.L. Stevens (Eds.), Mental models (pp. 155-190). Hillsdale, N.J.: Lawrence Erlbaum, 1983.
16. Forbus, K.D. & Stevens, A.L. : Using qualitative simulation to generate explanations. Report No. 4490. Cambridge, MA: BBN Laboratories, 1981.
17. Goldstein, I. : The genetic graph: a representation for the evolution of procedural knowledge. In D. Sleeman & J.S. Brown (Eds.), Intelligent tutoring systems (pp. 51-77). New York: Academic Press, 1982.
18. Johnson-Laird, P.N. : Mental models. Cambridge: Cambridge University Press, 1983.
19. Kluwe, R.H. & Haider, H. : Modelle zur internen Repräsentation komplexer technischer Systeme. Sprache & Kognition, 9, 173-192, 1990.
20. Langley, P., Ohlsson, S. & Sage, S. :. Machine learning approach to student modelling. Technical Report CMU-RI-TR-84-7. Pittsburgh, PA: Robotics Institute, Carnegie-Mellon University, 1984.
21. Mandl, H. & Lesgold, A. (Eds.) : Learning issues for intelligent tutoring systems. New York: Springer, 1988.
22. Ohlsson, S. & Langley, P. : Psychological evaluation of path hypothesis in cognitive diagnosis. In H. Mandl & A. Lesgold (Eds.), Learning issues for intelligent tutoring systems (pp. 42-62). New York: Springer, 1988.
23.Payne, S.J. : Methods and mental models in theories of cognitive skill. In J. Self (Ed.), Artificial intelligence and human learning (pp. 69-87). London: Champan and Hall, 1988.
24. Polson, M.C. & Richardson, J.J. : Foundations of Intelligent Tutoring Systems. Hillsdale, N.J.: Erlbaum, 1988.
25. Psotka, J., Massey, L.D. & Mutter, S.A. (Eds.). : Intelligent Tutoring Systems: Lessons learned. Hillsdale, N.J.: Erlbaum, 1988.
26. Reimann, P. : Towards general knowledge - diagnosis systems for student and user modeling. In H. Mandl, E. de Corte, N. Bennett & H.F. Friedrich (Eds.) Learning and Instruction. European Research in an International Context. Volume 2.1. Social and cognitive Aspects of Learning and Instruction (pp. 349-365). Oxford: Pergamon Press, 1990.
27. Reiser, B.V., Anderson, J.R. & Farrel, R.G. : Dynamic Student modeling in an intelligent tutor for LISP programming. Proceedings of Ninth International Joint Conference of Artificial Intelligence (pp. 8-14). Los Altos, CA: Morgan Kaufmann, 1985.
28. Rumelhart, D.E. & Norman, D.A. : Representation in memory. CHIP Report 116. San Diego: University of California, 1983.
29. Sauers, R. & Farrel, R. : GRAPES user's manual, Technical Report 1, Department of Psychology, Carnegie-Mellon University, Pittsburgh, PA, 1982.

30. Sleeman, D.H. : Inferring (mal) rules from pupils' protocols. Proceedings of the European Conference on Artificial Intelligence (pp. 160-164). Orsay, France, 1982.
31. Sleeman, D. & Brown, J.S. (Eds.) : Intelligent Tutoring Systems. New York: Academic Press, 1982.
32. Van Lehn, K. : Student modelling. In M.C. Polson & J.J. Richardson (Eds.), Foundations of Intelligent Tutoring Systems (pp. 55-78). Hillsdale, N.J.: Erlbaum, 1988.
33. White, B.Y. & Frederiksen, J.R. : Progressions of qualitative models as a foundation for intelligent learning environments. Report No. 6277. Cambridge, MA: BBN Laboratories Inc., 1986.
34. White, B.Y. & Frederiksen, J.R. : Qualitative models and intelligent learning environments. In R.W. Lawler & M. Yazdani (Eds.), Artificial Intelligence and education; Volume 1: Learning environments and tutoring systems (pp. 281-305). Norwood, N.J.: Ablex Publishing, 1987.
35. White, B.Y. & Frederiksen, J.R. : Causal model progressions as a foundation for intelligent learning environments. In W.J. Clancey & E. Soloway (Eds.), Artificial Intelligence. Special issue on artificial intelligence and learning environments (pp. 99-157). 42, 1. Amsterdam: North-Holland, 1990.

Computer-Based Learning Environment and Automatic Diagnosis System for Superposition of Motion

Aemilian Hron, Jürgen Bollwahn, Heinz Mandl, Uwe Oestermeier, and Sigmar-Olaf Tergan

Deutsches Institut für Fernstudien an der Universität Tübingen, Konrad-Adenauer Strasse 40-44, D-7400 Tübingen 1, FRG

Abstract: In the context of a computer-based learning environment on superposition of motion, an automatic diagnosis system on knowledge acquisition and on misconceptions was developed. The learning environment is designed according to the principle of inductive learning. It consists of a sequence of sixty similar tasks increasing in task demand and tutorial measures, involving simulations of the physical phenomenon. The diagnosis system is enabled to generate automatically both the correct solution and incorrect solutions which are based on the learner's possible misconceptions. It compares these solutions with the learner's actual solution and thus infers on his or her respective conceptual knowledge. The diagnosis system is rule-based and implemented as a classifier system. It uses both a strengthening algorithm and a discrimination algorithm. Learning environment and diagnosis system are implemented in LOOPS on a Xerox 1108 workstation.

Keywords: automatic diagnosis, causal reasoning, classifier system, learning environment , misconception, production system, qualitative modeling

1. Introduction

In the context of a research and development project and in connection with research on psychological issues on knowledge representation and knowledge acquisition, two components of an Intelligent Tutoring System intimately related with each other were developed: a computer-based learning environment on physical superposition of motion and an automatic diagnosis system on knowledge acquisition and misconceptions [2].

The research interests of the project are related to physical domains wherein imagery aspects of knowledge and everyday experiences are of import. The focus is on qualitative knowledge and on the approach on mental models (see the contribution of Mandl & Hron for more detail). Issues concerning the transition from ordinary physical knowledge to more abstract knowledge are pursued. The phenomenon of superposition of motion appears to be an appropriate research domain, because learners are quite knowledgeable about it and because it shows up narrow relations with everyday experience and commonsense knowledge.

NATO ASI Series, Vol. F 86
Intelligent Learning Environments
and Knowledge Acquisition in Physics
Edited by A. Tiberghien and H. Mandl

In the following, less the psychological analyses and findings are going to be presented but rather the realized developments, i.e. the learning environment and the automatic diagnosis system.

2. Computer-Based Learning Environment

The computer-based learning environment is devised according to the principle of "learning by doing" along the lines of inductive learning. It consists of a sequence of about sixty physical tasks on the composition and decomposition of superposing motions. The problems are similar to each other. However, they vary in some decisive parameters and activate knowledge transfer with respect to each consecutive task.

All of the tasks are based on a scenario wherein a mother and a child step on a huge conveyor belt at point A in a leisure park. The conveyor belt moves to the right-hand side. The mother always remains motionless on the conveyor belt after she has stepped on it. The child always runs away from the mother: either to the right in the conveyor belt's direction, to the left opposed to the conveyor belt's direction, orthogonally or obliquely to the opposite conveyor belt's side. The phenomenon of superposition of motion results of the child's motion on the conveyor belt in motion.

To illustrate this, an instance of a task on composition of superposing motions is presented (see Figure 1). In this task, the learner has to determine the child's point of arrival who is running to the right-hand side on the conveyor belt in motion. The learner is given the relation of velocities of the conveyor belt's motion and of the child's own motion as well as the mother's point of arrival. In order to be able to respond to the task, the learner has to grasp that the child's own motion and the conveyor belt's motion superpose. He or she has to determine the child's own motion and to assemble it with the conveyor belt's motion, in order to determine the complete motion and thus the child's point of arrival (point "E").

The variability of tasks in the learning environment is achieved in that the child runs away from mother into different directions, the relation of child's own motion to conveyor belt's motion is altered, either composition or decomposition of motions is required, the conveyor belt stops and the child continues running or in reverse. Furthermore, the tasks' respective question can focus on specific aspects, for example, on the point of arrival or the relation of velocities of the composite motions. By no means tasks are that simple to be responded to as might be suggested by the presented example. As a tutorial measure for each task, the learner can elicit a simulation of the respective motion, and, according to the diagnosed misconception, can receive specific instructions. In regard to the project's psychological issues, the variability of the tasks allows for sequencing of the tasks in view of the varying degrees of similarity; this in order to investigate the conditions of inductive learning in the physical domain.

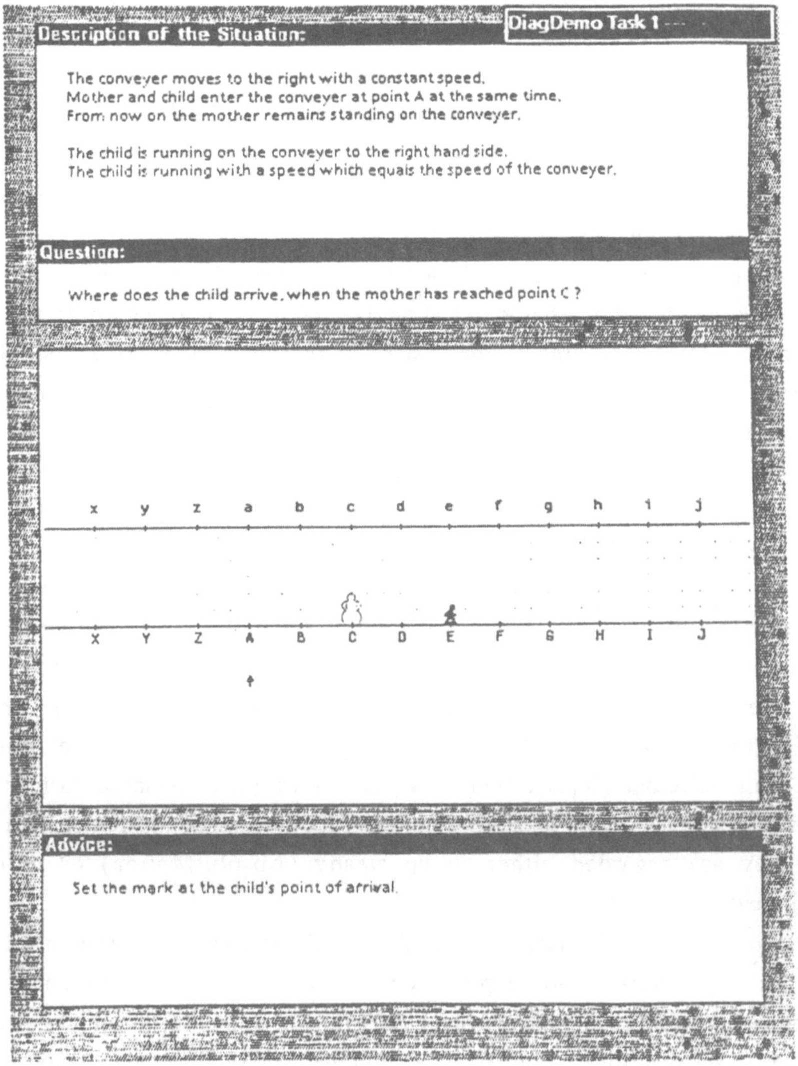

Fig. 1. Physical task: composition of superposing motion

3. Physical Concepts and Misconceptions

The physical knowledge which underlies this learning environment has been determined in the framework of cognitive task analyses. Two goal concepts are to be distinguished, which the learner is supposed to acquire, as well as seven misconceptions, which the learner can impose on the physical facts and which are to be overcome.

The two goal concepts are "relativity" and "independence". Relativity means that motions are always to be conceived in relation to a system of reference, in other words, that motions appear to be different according to the system of reference taken as a basis for them. Independence means that superposing motions do not influence each other, but ensue independently from each other.

Concerning the task shown above (see Fig. 1) the concept "relativity" means that for the child's own motion the conveyor belt (or the mother) serves as system of reference. The system of reference for the conveyor belt consists in the child's surroundings (or the firm row of letters). The learner has to take that into account. Furthermore, the learner - according to the principle of independence - has to regard the child's own motion and the conveyor belt's motion as independent from each other. That is, both motions do not influence each other.

Furthermore, empirical studies evidenced seven misconceptions: motions form a compromise (Compromise), one of the motions is predominant (MainMotion), motions are dependent from each other (Dependence), active motions are predominant compared to passive ones (ActivePassive), motions loose momentum (ModifiedImpetus), motions are absolute whereby they are regarded either to be steady (AbsolutMotion) or to diminish (ClassicalImpetus).

For the purpose of illustration we shall explain the effect of the misconception "Dependence" in the response to the presented task (see Fig. 1). In this misconception, the learner mistakenly assumes that the conveyor belt's motion influences the child's own motion. Therefore, he or she determines a longer distance of the child on the conveyor belt and a point of arrival which lies on the right-hand side of point E.

In case the child runs opposite to the conveyor belt's motion, that is to the left, the learner will also determine a wrong point of arrival. He or she assumes a reduced child's own velocity. In case the child runs away from mother perpendicularly to the conveyor belt, the learner assumes a modified direction of the child. He or she then assumes that the resulting oblique, total motion is going to be even more slanted.

4. Diagnosis System

4.1 Qualitative Modeling of Causal Reasoning

In the context of the learning environment, a diagnosis system was developed. The core of this system consists of a set of rules (or classifiers) by means of which conceptual as well as operational knowledge applicable for the correct and incorrect solution of tasks on superposition of motion is represented. An essential part of this set of rules are qualitative rules which reflect the learner's causal commonsense knowledge in the context of each of the concepts and misconceptions. In these rules, the learner's qualitative reasoning is described in terms of a qualitative representation of central parameters of a motion. In this case the central parameters are velocity and direction which we describe qualitatively as being constant, increasing, or decreasing. The modeling refers exclusively to changes of child's motion on the conveyor belt - in terms of modifications of the child's own motion and its direction.

For the purpose of illustration, the rules (or classifiers) for two misconceptions, "Dependence" and "ActivePassive", are presented in Figure 2. The first three rules describe the misconception "Dependence"; each rule standing for one of the three task situations: the child's own motion is directed either to the right, to the left or orthogonally. The misconception "ActivePassive" is described by the other three rules. In this instance as well, each rule refers to the child's respective direction.

The first one of the presented rules describes the effects of the misconception "Dependence" on the response of a task wherein the child runs to the right-hand side. This case was already presented in Fig. 1; the learner now assumes an influence of the conveyor belt's motion onto the child's own motion. The rule's condition part (IF) refers to the misconception ("Misconcept: Dependence") and the direction ("Direction: right") of the child's motion. The rule's action part (THEN) describes the learner's reasoning about the respective qualitative modifications of the child's own motion ("ChildVelocity: +") and its direction ("ChildDirection: ="). Thus the rule defines that, if the misconception "Dependence" prevails and if the child's motion is directed to the right, he or she will assume an increase in child's motion, but a constant child's direction. The next two rules on the misconception "Dependence" describe accordingly the learner's reasoning in case the child runs to the left (or orthogonally) to the conveyor belt's motion.

The next three rules describe the learner's qualitative reasoning on the basis of the misconception "ActivePassive". In this misconception, due to the child's running activity, the learner considers the child to be the more active element as compared to the conveyor belt. The first of these three rules ("ActivePassive-right") defines that in case of child's right-hand side motion (see action part), the learner assumes an increased child's own motion and a constant direction (see condition part). Note that the very same assumption is made by the learner, if the above-mentioned misconception "Dependence" prevails (see rule "Dependence-right"). Several overlappings of certain misconceptions occur in this way which, however, can be handled quite well by means of the diagnosis system (see below).

The next two rules on the misconception "ActivePassive" describe the learner's reasoning in case that the child runs to the left (or orthogonally) to the conveyor belt's motion. All of the misconceptions are being described in this way by means of qualitative rules.

```
(CLASSIFIER Diagnoser1 Dependence-right
        (If ((2 Misconcept:Dependence Direction:right))
            THEN
            (3 ChildVelocity:+ Childdirection:=))
        (Strength 157)
        (Discriminate NIL)
        (Trace T) )

(CLASSIFIER Diagnoser1 Dependence-left
        (IF ((2 Misconcept:Dependence Direction:left))
            THEN
            (3 ChildVelocity:- ChildDirection:=))
        (Strength 100))

(CLASSIFIER Diagnoser1 Dependence-ortho
        (IF ((2 Misconcept:Dependence Direction:ortho))
            THEN
            (3 ChildVelocity:= ChildDirection:- ))
        (Strength 100))

(CLASSIFIER Diagnoser1 ActivePassive-right
        (IF ((2 Misconcept:ActivePassive Direction:right))
            THEN
            (3 ChildVelocity:+ ChildDirection:=))
        (Strength 157))

(CLASSIFIER Diagnoser1 ActivePassive-left
        (IF ((2 Misconcept:ActivePassive Direction:left))
            THEN
            (3 ChildVelocity:+ ChildDirection:=))
        (Strength 100))

(CLASSIFIER Diagnoser1 ActivePassive-ortho
        ( IF ((2 Misconcept:ActivePassive Direction:ortho))
            THEN
            (3 ChildVelocity:+ ChildDirection:+))
        (Strength 100))
```

Fig.2. Rules for describing the misconception "Dependence" and "ActivePassive"

4.2 Diagnosis Network

Beside the qualitative rules described above, the diagnosis component consists of several other rules. The complete set of rules is presented in Fig. 3 in form of a so-called diagnosis network. The network structure results from the forward chaining of the rules. The nodes represent the rules, the links the potential follow-up relations between them.

According to the rules' different functions, the set of rules is subdivided into four layers (see the bottom line in Fig. 3): task classification, concept classification, qualitative assessment (qualitative rules), numerical computation. The rules of the first layer classify problems according to whether a composition or decomposition of motion is afforded. Rules of the second layer serve the purpose of identifying concepts and misconceptions and thus the monitoring of the application of specific tutorial measures. Rules of the third layer describe qualitative reasoning with regard to concepts and misconceptions; they have been illustrated above (see Fig. 2). Rules of the fourth layer are defined as effectors. In opposition to the other sets of rules, they do not produce any output to be applied within the rule system. Instead they activate LISP-functions for the calculation of results.

Each new task presented to the learner functions as input data for the rule-system and leads to the reconstruction of the diagnosis network. Thereby the structure of the network is changed, the rules, however, remain. In this way, for each new task the diagnosis network calculates on-line every possible incorrect solution as well as the correct one. Thereby the respective task conditions as well as the qualitative reasoning specific for the concepts and misconceptions are taken into account automatically. By means of comparing these solutions with the learner's task solution his or her concept knowledge is inferred.

4.3 Classifier Mechanism

The diagnosis network is implemented as a classifier system [3] by means of representing the rules as classifiers. A classifier system is a special form of production system that has specific computational advantages: speed and small memory requirements, easily implemented learning processes. Also typical for a classifier system is parallel firing of rules (classifiers). In opposition to traditional production systems, a classifier system uses a restricted symbolic notation: A classifier is written as a bit vector. The working memory of traditional production systems corresponds to the message list, which is a list of bit vectors. Our classifier system takes over the issue of bucket brigade algorithm for strengthening of rules and the issue of parallel firing. Instead of genetic algorithms, which are also typical of the classifier approach, we use a discrimination process.

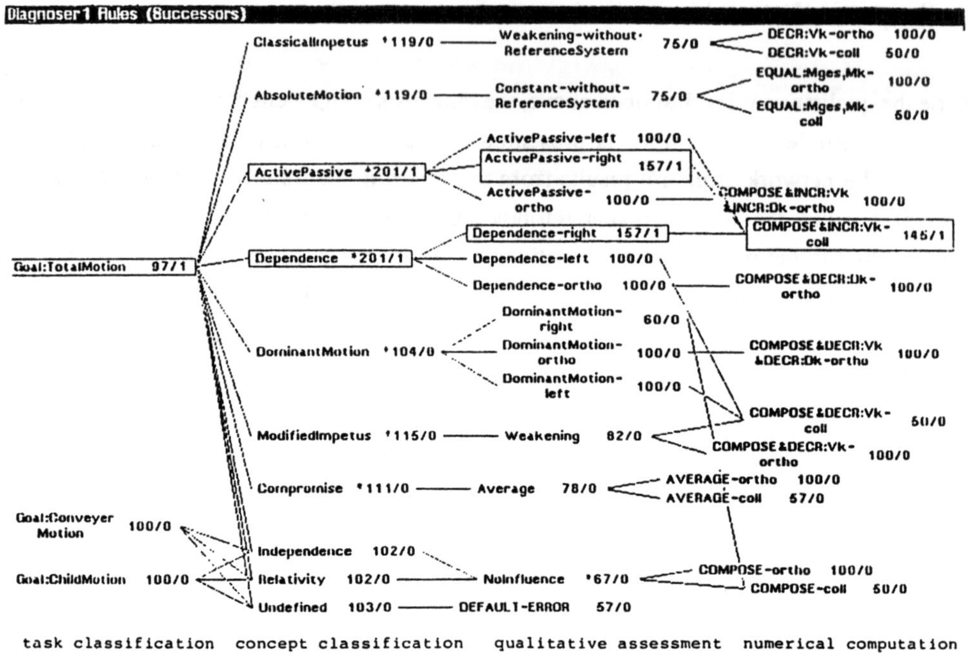

Fig. 3. Diagnosis network

For the presentation of the information in the classifiers (rules), the bit-vector presented in Fig. 4 is used which encompasses 39 positions. In this figure, the positions are occupied by so called don't care values (#). The don't care value is a special symbol used for pattern matching in a classifier system. The positions have different meanings. For example, the second to fourth position serve the encoding of information about the level of rules, the 21st to 33rd position the encoding of qualitative values, and so forth. The positions are then occupied with zeros, ones or don't care values (#).

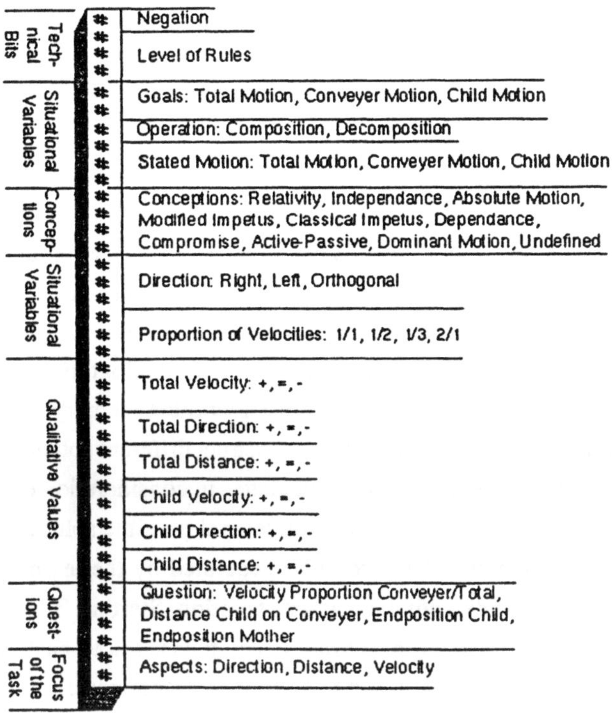

Fig.4. Bit vector

For the purpose of illustration in Fig. 5 a rule explained above on the misconception "Dependence" is shown concerning a task situation wherein the child moves to the right. In the upper part of Fig. 5 is the classifier-notation, in the lower part there is a directly readable description for illustration. The classifier consists both in the condition part and the action part of the described bit-vector. By means of zeros and ones at the respective positions, the respective information is encoded, which can be inferred from the readable version at the bottom.

```
(CLASSIFIER Diagnoser1 Dependence-right
          (IF ("#011######01101########################")
             THEN("#100#######################1001#########")
          (Strength 157)
          (Discriminate NIL)
          (Trace T))

(CLASSIFIER Diagnoser1 Dependence-right
          (IF ((2 Misconcept:Dependence Direction:right))
             THEN
             (3 ChildVelocity:+ ChildDirection:=))
          (Strength 157)
          (Discriminate NIL)
          (Trace T))
```

Fig. 5: Classifier

The system starts with an input vector on which the task characteristics are written. Then those classifiers fire that match with this input vector. Their action parts pose new information on this vector. This vector is pushed on a message list, which - in the next cycle - is matched again against the classifiers, and so on. This functioning is analogous to a traditional production system, wherein the productions match against a working memory.

On each layer, the classifiers fire quasi parallel and construct different solution paths which lead to the respective effectors. The effectors now calculate - in consideration of the different misconceptions and the respective task conditions - the several possible solutions to the task. These solutions, in turn, are compared to the solution generated by the learner. One of the effector-solutions happens to coincide with the learner's solution. Starting from that effector, the solution path to the corresponding concepts or misconceptions is traced back. Fig. 3 shows such a solution path which contains an overlapping of two misconceptions discussed above.

4.4 Strengthening and Discrimination of Rules

The solution paths of the diagnosis network, which lead to a solution corresponding to the learner's response, are stressed by means of strength values. The strength value is by and by reduced, if a rule (classifier) does not any more lie on a successful solution path. Tutorial measures can be initiated on the part of the learning system by means of threshold-value comparisons.

In the discrimination, task conditions, which are probably relevant for the observed solutions, are adopted within additional rules. Thereby the diagnosis shall adapt the rules to differing learning behaviour. First, the diagnosis system starts with rules formulated in general terms which are specified in the course of the session.

4.5 Automatic Selection of Special Tasks for a More Efficient Diagnosis

Since in many tasks different solution paths lead to the very same solution, it is not warranted that a misconception can be identified unambiguously. Such a case is presented in Fig. 6 in regard of the known task in Fig. 1. If the learner provides an incorrect response, for example, "EFE" (child's point of arrival is between E and F, but closer to E), then this can be attributed both to the misconception "ActivePassive" and "Dependence". The overlapping of these two misconceptions has been illustrated below.

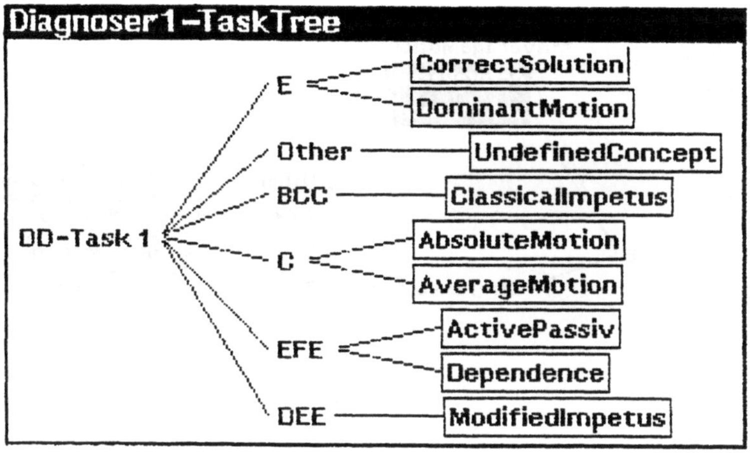

Fig. 6. Solution-overlapping

The solution paths overlap with respect to the solutions, which, however, vary at every task. Therefore, in the course of the session, different strength values nevertheless arise with regard to the misconception descriptive rules. This process can be enhanced by means of selecting the tasks precisely as to produce different overlappings [1]. For this purpose, a table for each task is elaborated by means of the diagnosis network and the calculation-component, in which the resulting solution is calculated and recorded for every misconception.

By means of the misconceptions-solutions tables elaborated by the system for each task, it is now straightforward to find a task wherein the solutions for the relevant misconceptions can be distinguished. Fig. 7 shows a selected successor task ("DD-Task 3") concerning the child running to the left. This task has differing solutions with regard to the misconceptions "ActivePassive" and "Dependence"; it is more suitable for the assessment of these specific misconceptions than the previous task.

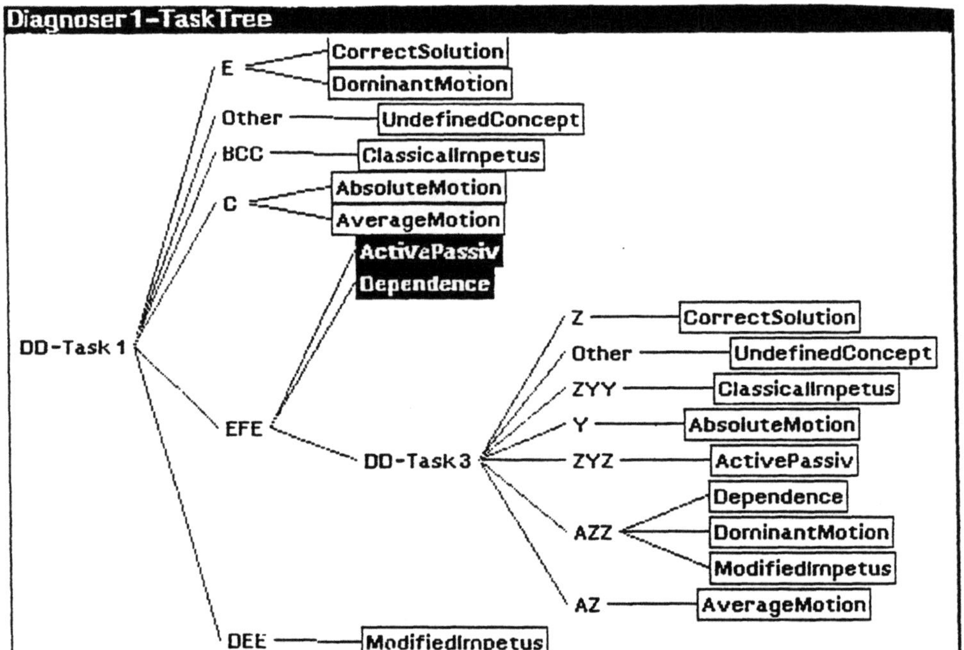

Fig. 7. Selected successor task

By means of the directed task selection the differing level of strength values is speeded up and thus the diagnosis as a whole becomes more efficient. That is, less task presentations are necessary to identify a misconception and to activate a corresponding tutorial measure.

5. Final Remarks

The presented learning environment offers appropriate opportunities for the acquisition of physical concepts on superposition of motion by means of inductive learning. Its basis are the different tasks within a uniform task scenario which foster transfer of knowledge onto each next task. Since one could vary the task sequences systematically according to the respective similarity relations, the conditions of inductive learning can be investigated experimentally in this physical domain. At the moment the project work is dedicated to this research.

The automatic diagnosis system shows up narrow links with the conception of the learning environment. Its special use for further applications lies in its suitability for computer-based learning environments in the sense of microworlds. In such learning environments, the learner is supposed to experiment freely and compose the task sequences by him or herself. Thus the task sequence can not be anticipated fully. This affords diagnosis systems which are characterized by high flexibility and which are enabled to generate solutions in the face of different task conditions and knowledge concepts. The presented diagnosis system possesses these features. It is an appropriate diagnostic instrument to correspond appropriately to the demands of automatic knowledge diagnosis in microworlds.

References

1. Becker, L.A.: Computer-Aided Misconception-Based Intelligent Tutoring and Exercise Generation. In: Programmed Learning and Educational Technology - Journal of the Association for Educational & Training Technology, Vol. 25 (pp. 67-75). Kogan Page 1988
2. Bollwahn, J., Hron, A., Mandl, H., Oestermeier, U. & Tergan S.-O.: Repräsentation von Wissen und Wissenserwerbsprozessen auf der Grundlage von Classifier-Systemen. In: Abstracts des III. Workshop "Intelligente Lernsysteme" (R. Gunzenhäuser & H. Mandl, Hrsg.), S. 25-40, 8.-9.6. 1989, Tübingen.
3. Holland, J.H.: Escaping Brittleness: The Possibilities of General Machine Learning Algorithms Applied to Parallel Rule-Based Systems. In: Machine Learning: An Artificial Intelligence Approach (R.S. Michalski, J.G. Carbonell, & T.M. Mitchell, eds.), Vol. 2, pp. 539-623. Los Altos, California: Kaufman 1986.

5. Final Remarks

The presented learning environment offers appropriate opportunities for the acquisition of physical [illegible] on superposition of motion by means of inductive learning [illegible] different tasks within a uniform task scenario which fosters transfer of knowledge into each other. [illegible] task sequences systematically [illegible]

Eliciting Hypothesis-Driven Learning in a Computer-Based Discovery Environment

Peter Reimann

Department of Psychology, University of Freiburg, Niemensstr.10, D-7800 Freiburg, FRG

Abstract: A computer-based interactive simulation environment is described that focuses on a specific type of inductive learning: learning by hypothesis testing. The program named REFRACT puts students in a situation where they can learn about a domain that resembles geometrical optics by means of scientific discovery learning: by designing experiments, stating predictions, and analyzing results. The students interaction with the program is organized in a way that provides both structure to students' task and a window on their hypothesis generation and testing behavior. The system adapts in some sense to students' knowledge by permitting them to state predictions in several ways, ranging from purely qualitative/graphical to quantitative.

Keywords: simulation, inductive learning, hypothesis testing, physics learning

1. Introduction

Forming and testing hypotheses using observations and empirical findings is not only central to scientific work, but also to problem solving and the acquisition of knowledge in general (24). The importance of forming hypotheses in the process of acquiring knowledge has in particular been stressed in the context of instructional approaches which focus on the active role of the learner. Bruner's discovery learning theory (4), Schuman's Inquiry Training (34), the method of Socratic Tutoring (6), or Palinscar and Brown's Reciprocal Teaching Program (18) are prominent examples. Brown, Collins and Harris (3) see hypothesis-driven learning as fundamental for understanding in many areas of learning and show examples from mathematics, physics and reading. In short, from research on cognition and motivation, we can conjecture that a hypothesis-guided approach to learning has effects on three levels: (a) the acquired knowledge is well structured and flexible to apply (cognitive effects); (b) students become aware of strategies useful to search for information actively (metacognitive effects); (c) hypothesis-guided learning is intrinsically motivating and raises self-esteem.

Hypotheses play a particularly important role in the course of scientific discovery. Many questions pertinent to complex induction tasks such as scientific discovery can effectively be studied by confronting students with microworlds. In these environments, students have to engage in activities such as conducting (simulated or real) experiments, recording data, and generalizing and abstracting from specific instances. The student is seen as an active learner who explores interesting phenomena with the goal of a deeper understanding of the domain.

NATO ASI Series, Vol. F 86
Intelligent Learning Environments
and Knowledge Acquisition in Physics
Edited by A. Tiberghien and H. Mandl

Several authors have proposed to use microworlds as an instructional tool to foster learning in mathematics and the sciences (e.g. [7, 21, 39, 41]). One expectation is that microworld-based instruction will help overcome the often quite stable misconceptions found in students even after many hours of formal instruction [5, 14]. Carefully designed microworlds should allow students to acquire a more meaningful understanding of scientific principles and of the nature of science itself [28, 41].

Existing microworlds mostly address misconceptions developed by students in domains such as mechanics [5, 30] and electronics [31, 32]. For example, diSessa [5] and White [30] recognized that students have an Aristotelian "theory" of particle motion even after many hours of training in Newtonian mechanics. They describe computerized microworld programs that attempt to gradually change students' misconceptions by confronting them with situations incompatible with Aristotelian physics.

In my research, I build on the microworld paradigm, focusing not as much on misconceptions about a domain but on the processes involved in discovery learning, that is, on inductive inquiry skills. This refers to students' effectiveness in collecting, organizing, and understanding data, concepts, and relationships in a new domain. To emphasize this focus, I talk about *discovery worlds* as distinct from other types of microworlds. In this chapter, a discovery world program that builds on the domain of geometrical optics is described. The intention behind the construction of this program was not primarily to teach to students a specific subject area or to help them overcome misconceptions about a specific phenomenon, but to provide them with experiences in applying certain scientific discovery skills and thereby study the psychological processes involved in acquiring and executing these skills. Processes involved in complex induction tasks such as scientific discovery have been analyzed by psychologists to some extent (for an overview see [9, 35]). However, in these studies components of the discovery process are usually analyzed in isolation. There are studies on experiment construction [11, 17, 29], hypothesis testing [36, 37], and analysis of results [27, 38] with only few interrelations. Furthermore, the central aspect of discovery - generation of hypotheses and theories - has hardly been analyzed at all. In the last years, however, research within the information-processing paradigm of cognition in combination with results from the Machine Learning area (e.g., [15, 16]) has helped to shed more light on the processes involved in generating hypotheses and their interaction with activities of data gathering and hypothesis testing [8, 9, 12, 13, 30].

2. The Discovery Environment REFRACT

REFRACT is an interactive simulation program that has been designed with the intention to elicit in students processes of hypothesis formation and testing. Furthermore, the environment should make it possible to observe the students hypothesis formation behavior in as much detail as possible. In order to satisfy these goals and to have a motivating task for the student, I decided to construct a computer program in which the student has to behave in a way that resembles the process of (empirical) scientific discovery, although in a simplified and constrained manner. The program has the form of a simulated physics laboratory in which the student can run experiments in order to find out about the relations between certain variables. The variables are connected to objects that play a role in geometrical optics: surfaces, lenses, rays. Examples for variables are: optical density, angles, distances. The relations between the variables resemble certain laws of geometrical optics: ***Snell's Law*** and the ***Lens Maker's Equation***. While learning about the functional relations between the variables, the student has the opportunity to engage in science-related activities such as conducting experiments, recording and analyzing data, and formulating and testing hypotheses. Since the program is primarily used as a research tool, it keeps detailed records of students' actions.

The objects and the regularities realized in REFRACT only *resemble* those relevant for geometrical optics since REFRACT does not actually incorporate the physically correct laws. Instead, simplified relations are used. The reason for this is that the law of refraction (Snell's Law) - which describes the phenomenon that light rays "bend" when they go from one optical medium (such as air) into another (such as water) with different optical density - includes in its correct form trigonometric expressions:

```
sin(Angle of Incidence) / sin(Angle of refraction) = n21,
```

where n_{21} is a constant called the index of refraction of medium 2 with respect to medium 1. Since students working with REFRACT have to induce relations between variables from analyzing experimental data, detecting relations that incorporate trigonometric terms does not seem to be feasible for mathematically unsophisticated students. Therefore, a simplified version of Snell's law is used:

```
Angle of Incidence / Angle of Refraction = n21,
```

a version also used by physicists as an approximation to Snell's law when angles are small. Further simplifications that make the relations in REFRACT not cohere to the physical theory are the fact that idealized thin ("laser") rays are used so that there is no reflection occurring and that the values for the optical density of the media involved (*glass*, *flintglass*, *diamond*) do not correspond to their physical counterparts.

In other words, REFRACT should not be taken as a simulation tool for geometrical optics. Geometrical optics is only the semantic embedding meant to make the task more meaningful and motivating for the student. It is an computer-based environment where students can exercise certain skills that belong to the realm of scientific inquiry and where detailed traces of students behavior can be gathered. A learner in REFRACT is confronted with the problem to find out which regularities govern the behavior of the objects in the environment. Starting from single observations, her task is to infer more general assertions. More specifically, the following features characterize the learning task in REFRACT: (a) the learner is assumed to be a 'novice', to have no knowledge about refraction phenomena except for some rough starting hypotheses such as "Light rays bend when they go from air into another medium", or wrong knowledge; (b) the only sources of information for the learner are the outcomes of simulated experiments. These experiments have to be designed by the student herself; (c) the environment contains both relevant and irrelevant information. It is up to the student to find out which variables convey the relevant information; (d) the task is solved when the learner is able to predict the path of rays correctly for all possible combinations of light rays and media. The student does not have to explain *why* things happen the way they do; her only concern is with *predictive* power; (e) since the data are acquired in an incremental fashion by running one experiment at a time and since the learner has to make predictions based on incomplete information, she often will have to revise hypotheses based on new evidence.

In order to learn about "refraction" phenomena, the student can design experiments, observe what happens and store data in a computer simulated notebook. After designing an experiment and before seeing the result the student has to predict its outcome. Note that in each experiment a prediction has to be formulated (about the path of rays) before feedback can be received.

Phrased in more psychological terms, students' activities are organized according to what we call a SPFP sequence [32]: *Search* for information by conducting experiments and making measurements; *Prediction* of the experimental outcome; *Feedback* on the outcome; and further *Processing* of feedback information. SPFP sequences constitute an experimental setting that is particularly well suited for studying active discovery learning. First of all, it gives structure to students' interaction with the system. Secondly, the SPFP sequence is a useful strategy to assess the process aspects of hypothesis generation and testing behavior. The sub-components of this process - data selection, hypothesis/prediction formation, and hypothesis revision - are separated from each other without seriously disturbing the natural path of inductive learning. The computer can easily keep track of students' activities by recording the screen actions. Furthermore, computer records can be enriched by prompting the student for verbal comments.

2.1. The Hypothesis-Generation and Testing Cycle

The basic cycle of experimentation in REFRACT consists of three main steps: designing an experiment, predicting its outcome, and processing feedback. During the course of the learning session, the student can conduct whatever experiments she wants to pursue, within the limitations of the program. The ultimate goal for her is to find out as much as possible about the laws governing the phenomena.

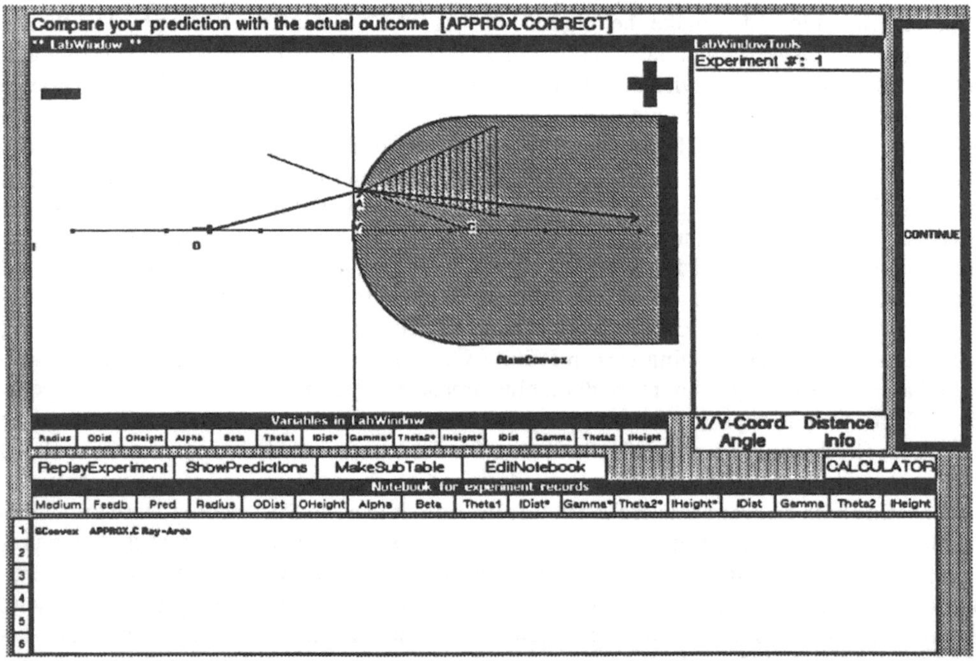

Fig. 1. REFRACT's main user interface

Figure 1 depicts the main screen of the program[2] with two central windows. In the *Laboratory Window* (abbreviated as *LabWindow*) experiments can be simulated in graphical form. The *NoteBook* window can be used to keep track of information about experiments and to manipulate numerical data

[2] REFRACT runs on Xerox 11xx Lisp Machines and is written in InterLisp and Loops. Loops is an experimental software product of Xerox that adds object-oriented programming capabilities to InterLisp.

The *Laboratory Window* is connected to several supporting windows which will be explained later. At the top of the screen is the general prompt window (shaded darkly); it is used to communicate general information about the state of the discovery world to the student.
Let me now trace the system through an experimental cycle, i.e., through the design, prediction and feedback step. I will use the single surface/single ray phenomena for purposes of illustration and not further discuss the second area available in REFRACT: image formation on lenses (see [24] for a complete description of the program).
Designing Experiments. In the design step, the student has to combine an optical medium with a ray. The student is guided through this phase by a sequence of pop-up menus. First, a medium has to be selected by choosing one of the alternatives from the medium type menu (Figure 2). The menu is ordered vertically according to surface shape and horizontally by substance. Differences in substance type are indicated as differences in the medium's degree of shading. If the student selects a convex or concave medium, there are each three radii available to choose from.

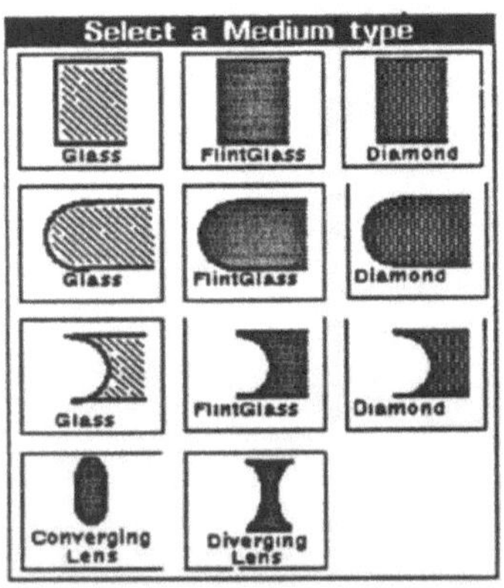

Fig. 2. Menu with the different media available for selection.

Next, the student can fix the distance between the medium and the object (measured with respect to the vertex point *V* of the object). This is done again by selecting from a pop-up menu. The object (an icon symbolizing a "flashlight" source of a single ray) is then drawn onto the screen at the point specified. All distances and coordinates are specified in terms of screen units. Then the student defines the input ray in terms of *Alpha*, the angle it forms with the *CenterRay*, a help line provided by the program. Six different values are offered, three for rays that go upwards from the *CenterRay* and three that go downwards. After the student has selected a value for *Alpha*, the program draws the input ray on the screen; also, the *Normal* at the point where the ray intersects the surface is drawn and labels are affixed for the points *O*, *A* and *V*. In Figure 3, a complete design for an experiment is shown.

To summarize the design step: the student has control over four variables which comprise the independent variables of the simulation: (1) the substance of the medium, which is varied nominally; possible values are *Glass*, *FlintGlass*, and *Diamond*; (2) the shape of the medium's surface, i.e., its radius; possible values are *plane*, *convex*, or *concave*; (3) the distance from the light source, *ODist*, with five values; (4) the angle the incoming ray forms with the *CenterRay*: *Alpha* (six possible values). Note that optical density is not represented directly by the system as an independent variable, but indirectly by means of differences in substance type. The student is supposed to find out that there exists some property of the medium (optical density) that affects the degree to which light rays are refracted. In other words, optical density is realized in the simulation as an intervening variable, one that cannot be controlled experimentally but affects the value of the dependent variables, *angle of refraction* and *image distance*.

Another design decision was to vary Radius, *ODist*, *Alpha* and *OHeight* discretely instead of allowing the student to select from a continuum of values. This was done in order to make it easier for the student to repeat design selections and to speed up the assembly of a design: selecting values from a small menu can be done much faster than to try to control variables by drawing rays on the screen..

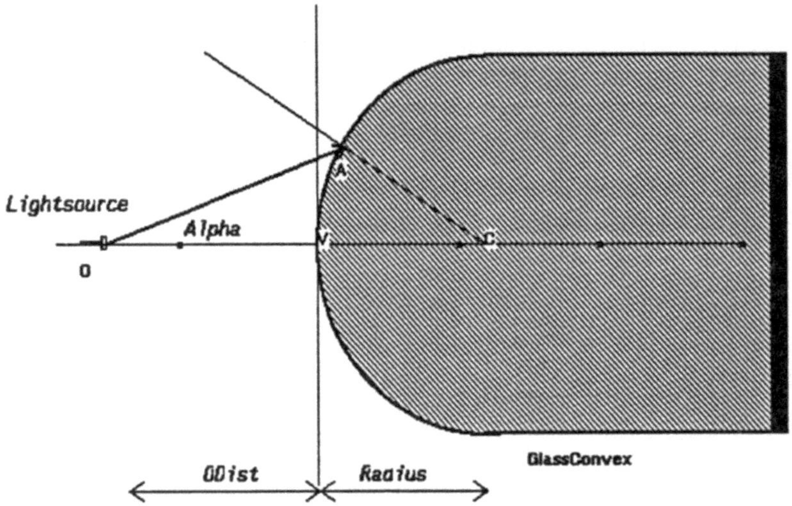

Fig. 3. The Laboratory Window after the design step is completed. The labels written in italic are not part of the actual display but are added to the figure for clarity.

Predicting the Ray Path. In the prediction phase, the path of the refracted ray has to be specified by the student. Several options are available to perform this task. The degree of a prediction's precision can be varied according to the student's knowledge and preference. The possibility to adapt the prediction's precision has two advantages. Firstly, we can draw conclusions from the prediction format to the quality of the hypothesis which stands behind it, and, secondly, we avoid frustrating the student by asking more from her than she can accomplish.

Predictions can be stated in a numerical or graphical manner. Graphical predictions are seen as being less precise than numerical ones. However, even within the graphical prediction types the degree of precision can be varied. The student can decide whether she wants to draw a single ray or to draw an area within which the refracted ray is presumed to lie.

In the case of numerical predictions, the student is simply asked to input numbers for all the necessary aspects of a prediction, such as angle of refraction (*Theta2**) or the image distance (*IDist**). As a convention, predicted values are distinguished from actual values by attaching a star ('*') to the former. For instance, *IDist** is the predicted image distance, *IDist* the

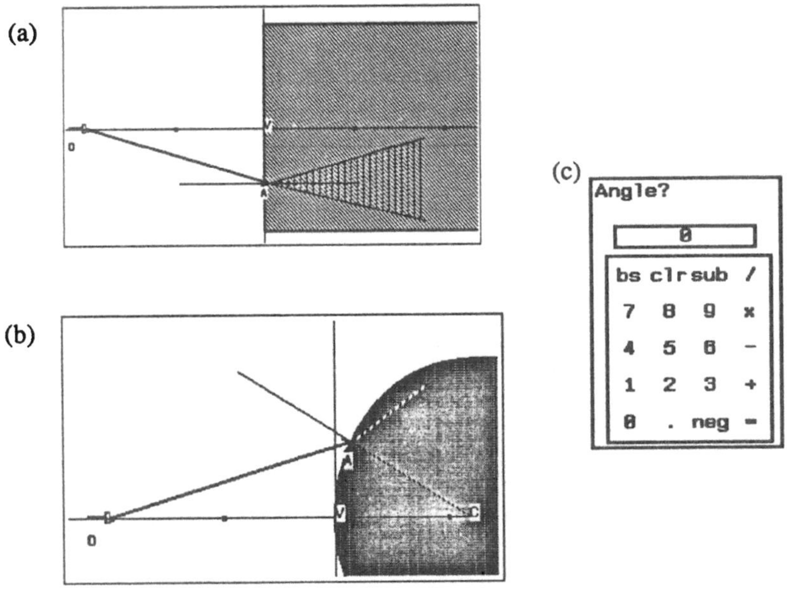

Fig. 4. Various ways to state predictions: (a) area prediction; (b) ray prediction; (c) number prediction

actual one. Figure 4 shows some of the possibilities for stating predictions numerically and graphically.

The student can enter a value for the image distance *IDist* or the angle of refraction *Theta2* using touchpad menus. Besides entering predictions graphically or numerically, REFRACT also allows for stating them in a form that is closer to the hypotheses that stay behind specific predictions. The idea is that the student enters a hypothesis which is turned into a prediction by REFRACT itself. However, the way a hypothesis can be entered directly is quite limited. The student can enter hypotheses only in the form of a statement comprising three components: a dependent variable, a relator, and a combination of independent variables. This statement is entered under the control of a group of menus shown in Figure 5. The menus pertain to options about dependent variables, independent variables, relators, and mathematical operators that can be used to combine independent variables. Relators can be exact ("=") or less exact ("Less Than", etc.). The independent variables combination can be either (a) a single variable (e.g., "Theta2"), (b) a constant, (c) a combination of constant and variable(s) (e.g., ".4 * Alpha"), or (d) a combination of variables ("Alpha/2 + Theta"). The student is guided through these menus by shading; only not shaded fields are admissible as the next entry. When a hypothesis is completely specified, REFRACT attempts to evaluate the term to the right of the relator by first binding the independent variables occurring in it to their current value and then evaluating the expression in order to yield a number. In case the student used the equal sign as relator, the calculated value will be utilized immediately to draw the predicted ray into the *Laboratory Window*. For example, if the student entered "Theta2 = Alpha/2 * 3.0", the specific value for *Theta2* is calculated and the prediction is executed in the *Laboratory Window*. However, if the student only expressed a constraint such as "Theta2 GT Theta1" (*Theta2* is greater than *The-*

tal), REFRACT uses this circumscription to draw a constraining ray in the *Laboratory Window* and then asks the student to complete her prediction by means of a numerical or graphical prediction format, as described above. In any case, the verbal formulation of a hypothesis is stored away by REFRACT and is accessible by the student at any time from the *NoteBook* (discussed below).

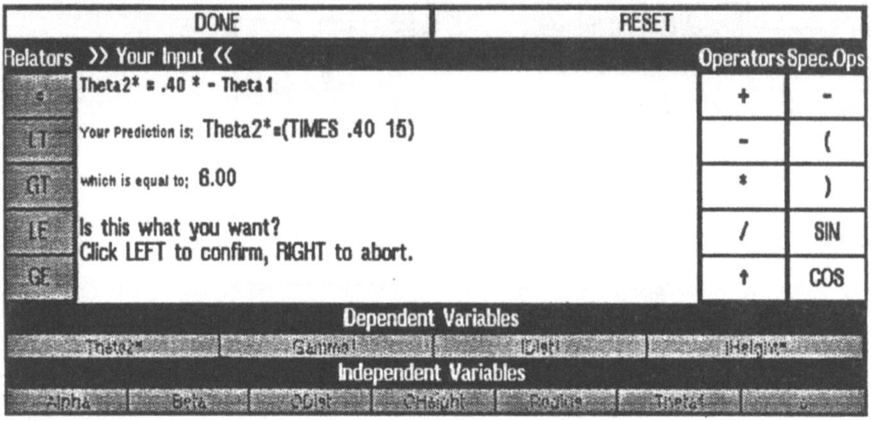

Fig. 5. Menu used to enter a hypothesis "verbally"

Receiving Feedback. Feedback is given in two forms: graphical and verbal. The correct ray is drawn into the *Laboratory Window.* Correct rays have arrowheads to distinguish them from predicted ones (see Figure 6). Verbal feedback is given in the form of a short qualification of the prediction. One of three values is assigned: *Correct, Approximately Correct*, or *Wrong.* This feedback is primarily provided to disambiguate the graphical feedback: Graphically small differences between predicted and correct ray path sometimes constitute major mistakes in terms of the underlying physical principles. Verbal feedback is given according to the following criteria. If the prediction is not in a physically reasonable area, the student receives the feedback "wrong"; if the prediction is in the correct area, but not as precise as possible, the prediction is qualified as approximately correct;otherwise it is said to be correct.
This concludes the description of an experimental cycle. It should be mentioned that the design and prediction step can be "undone"; after finishing each of these steps, the student is asked to confirm her decisions and, if not satisfied, she is put back to the beginning of the step. I will now introduce those aspects of the user interface that help the student to gather and analyze more information about experiments.

Tools for interrogating the simulation. The student sees initially only graphical information about an experiment. In order to inspect numerical aspects of an experiment she can use two menus which are located at the bottom of the LabWindow (see Figure 1).

Recording Tool. The *NoteBook* is a tool for the student to keep track of the sequence of experiments she conducts. Further, it provides options for organizing the information in a flexible manner and performing numerical operations on experiments. Figure 7 shows the *NoteBook* with some entries. The data are organized vertically according to the sequence of experiments and horizontally according to design, prediction and feedback variables. By means of the menu at the top of the *NoteBook* the student can, besides other things, edit the *NoteBook*, or replay a previous experiment.

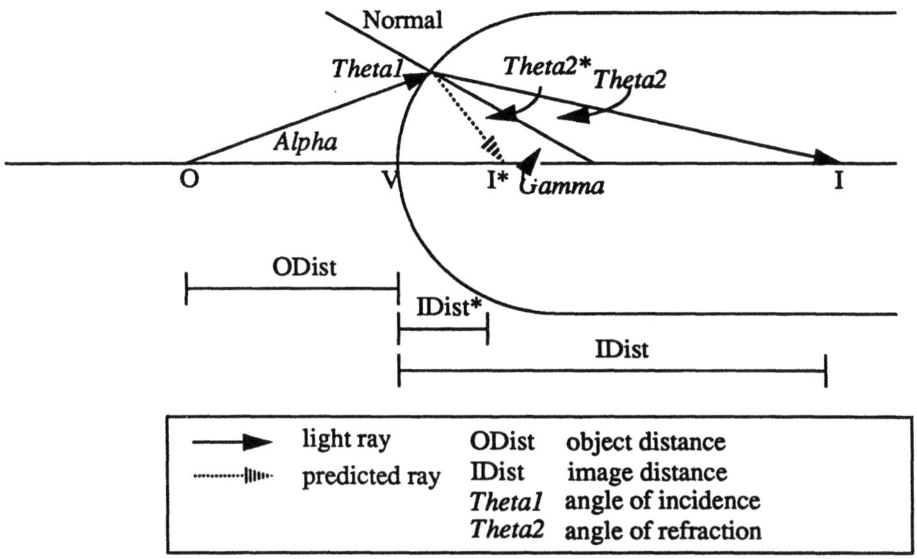

Fig. 6. Schematic display of feedback information.

The main function of the *NoteBook* is to display numerical aspects of experiments. This can be done by editing the *NoteBook*, i.e., by inserting values for variables into it. Information about experiments other than the current one may also be inserted at any time. Note that the program automatically inserts values for the first three columns of the *NoteBook*: Medium, Verbal Feedback, and Prediction Type used. All the other fields can be filled in by the student.

An experiment can be *replayed* after the feedback for the current experiment is produced and before a new experiment is started. The selected experiment will be replayed in the *Laboratory Window.*

Data Analysis Tool. In order to select, organize and manipulate information from the *NoteBook*, the student has the option to work on data presented in a tabular format. The student can select a subset from the variables which appear in the *NoteBook* and display them in a table (see Figure 8). Starting from this table, the student has the following options: (1) She can sort the values of a variable column; (2) she can select experiments with specific values from any column (except the experiment number and calculation variable column); (3) she can enter an equation that gets calculated for all experiments in the table. In Figure 8, a table is displayed together with a request to calculate *Theta2/Alpha* for all experiments in the table. The resulting values are displayed in the right-most column.

To summarize, the key features of the implementation are:

- The program provides ample opportunity for the learner to steer the process of forming and testing hypotheses through simulated experiments.

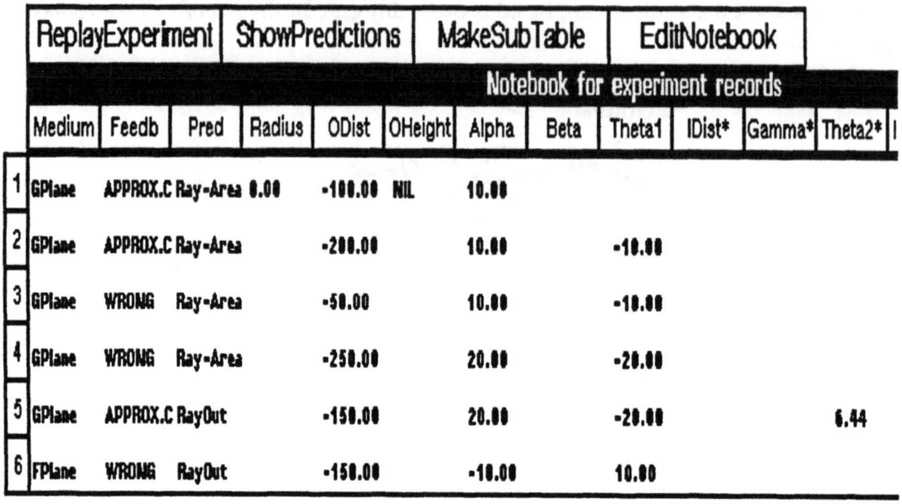

ReplayExperiment | ShowPredictions | MakeSubTable | EditNotebook

Notebook for experiment records

	Medium	Feedb	Pred	Radius	ODist	OHeight	Alpha	Beta	Theta1	IDist*	Gamma*	Theta2*
1	GPlane	APPROX.C	Ray-Area	0.00	-100.00	NIL	10.00					
2	GPlane	APPROX.C	Ray-Area		-200.00		10.00		-10.00			
3	GPlane	WRONG	Ray-Area		-50.00		10.00		-10.00			
4	GPlane	WRONG	Ray-Area		-250.00		20.00		-20.00			
5	GPlane	APPROX.C	RayOut		-150.00		20.00		-20.00			6.44
6	FPlane	WRONG	RayOut		-150.00		-10.00		10.00			

Fig. 7. The *NoteBook*

- The phenomena the student can learn about are presented in idealized form: The numerical relations are simplified and extreme refraction indices are chosen to make the refraction phenomena visually more salient.
- The process of information selection, hypothesis formation and hypothesis testing is sequenced into separate steps, thus allowing for a detailed assessment of the student's actions.
- Predictions about the phenomena can be made on different levels of precision according to the student's knowledge about the domain.
- Information about experiments in the learning environment is given in graphical and numerical form.
- Tools are given to the student to acquire more information about one experiment and to keep track of a whole series of experiments.
- Another set of tools provide the student with the opportunity to select, organize and manipulate (numerical) information.

2.2. Discussion

Experiences with the use of REFRACT. A small exploratory study was conducted wherein I observed eight subjects working with REFRACT for a couple of hours each (see [24]). Subjects (Ss) in this small study showed many differences, both with respect to knowledge acquired and to learning behavior. This is not surprising, given the openness of the learning situation. While almost all Ss found out that medium substance affects refraction, only a few of them could express this in terms of functional relations between dependent and independent variables. With respect to the learning process, I noticed differences with respect to control of variables, prediction making, discrimination of relevance, and determination of covariation. The latter relates to Ss skill in keeping track of experimental observations and finding regularities in data. Finally, Ss felt more or less comfortable with the guidance-free learning situation.

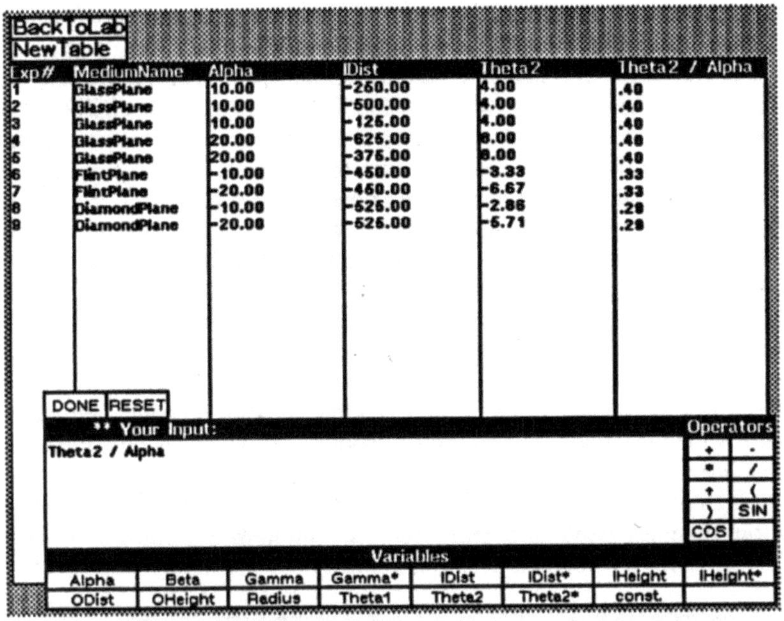

Fig. 8. Example for the use of a table to calculate an expression for all experiments stored in the *NoteBook*

I think that two findings are of foremost importance. First, many Ss were not able or not willing to formulate predictions which are precise in the sense that they are based on numerical relationships between variables. This is in part reflected by the fact that the tools in REFRACT which can be used to discover such relationships were not used very often. Closely related to this first point is a second central observation: the frequent use of graphical (pictorial) information. I did not expect that Ss would use this kind of information so often and constantly to build generalizations and to generate predictions.

Improving REFRACT for teaching inquiry skills. From the description of the components of the discovery task and from the observations on human learners it can be concluded that REFRACT provides an research tool that allows to invoke processes on side of the learner as they are central to the process of forming descriptive generalizations: experiment generation, hypothesis formulation, prediction making, hypothesis testing by comparing predictions to feedback. As such, REFRACT can be used to elicit behavior that is relevant for the study of (scientific) discovery. However, REFRACT in itself is not provided with means to foster the process of discovery learning. For example, it does not incorporate diagnostic tools that would help it to build a model of students knowledge and discovery skills nor does it have the means to utilize such a learner model for adaptive help or coaching activities.

In work reported elsewhere [25] I undertook a more detailed task analysis of the discovery problem in REFRACT, based on the problem solving model of induction and on a protocol analysis of learners' knowledge acquisition process. This task analysis led to the implementation of several computer simulation models using the production system formalism that attempt to explicate important characteristics of scientific discovery as exhibited by students working with REFRACT. Since these models of discovery are implemented as computer programs, they can be made directly useful for computer-provided instruction by serving as the basis of the diagnostic component of an Intelligent Tutoring System, i.e., the component that constructs automatically a model of the student's knowledge and skill.

But even without the development of a student modeling component, the program could be enhanced in order to focus the learner on improving her scientific inquiry skills. In REFRACT, the learner does not only produce experiments (data), but also predictions and under certain circumstances even hypotheses. While the experimental data and the predictions are explicitly represented in REFRACT (as graphical experiments and numerical data in tables), the representation of the more abstract but more fundamental hypothesis is underdeveloped: they are currently only represented in form of text strings the learner has to enter herself. One way of enhancing the display of hypotheses could be to present to the learner in graphical form the relation between hypotheses and evidence (experiments), for example by graphing such relations as "Hypothesis h covers experiments e_i" or "Experiment e contradicts hypotheses h_k". Such relations between theory and evidence are essential for driving the process of scientific discovery [10, 30]. I expect such devices for the *reification* of theory-evidence links to be instructionally effective to the degree that they (a) guide the student's attention, (b) lower the cognitive demands of the theory-to-evidence comparison (i.e., serve as an external memory), and (c) help the student to see differences between the current learning state and the goal state. This idea is quite similar to reification strategies used in problem solving situations as illustrated by Anderson's Geometry tutor [1] and Brown's Algebraland program [2].

Comparison with other work. The work most directly related to REFRACT concerns computer-based learning environments that build on the scientific discovery metaphor. One example is DIBI, a microworld program for the conservation of momentum [33]. In a manner quite similar to REFRACT, learners in DIBI have to generate experiments and predictions about a physics phenomenon. The domain simulation of DIBI is more advanced than the one of REFRACT in that it allows animated graphics to be displayed to the user. Furthermore, DIBI has first modules integrated for student modeling and system-guided experiment selection [19, 20] as well as a module that supports knowledge acquisition by means of analogical reminding [26].

At the Learning Research and Development Center, Glaser and co-workers have been developing and evaluating computer programs that besides conveying domain knowledge (about electricity: *Voltaville*, and economics: *Smithtown*) are also intended to actively coach students on their inquiry skills [22]. Rather than focussing on the construction of highly individualized student models based on students' hypothesis testing behavior, *Voltaville* and *Smithtown* incorporate a set of static critics that monitor students inquiry behavior and coach her on specific mistakes. In *Voltaville* and *Smithtown*, students can also entertain an observational strategy, that is, they are not forced to actively state predictions. While this is a quite natural strategy, in particular in the initial learning phase, it makes it harder to assess on-line what precisely it is the student is learning from her observations.

The discovery of descriptive generalizations which describe how entities change depending on changes of other entities is only one aspect of scientific discovery. What usually follows (or often comes first) is the formulation of explanatory generalizations, i.e., statements describing *why* things happen the way they do. In the natural sciences, explanatory generalizations are mostly formulated in terms of causal relations between the entities of interest. For example, in mechanics one not only notes the relation between certain descriptors ("as mass increases, acceleration decreases") but also explains these relations in terms of forces. New developments in educational computing attempt to provide student with means to in a sense discover such explanatory concepts in a manner that would not be possible without the use of computers. The main principle in this work is to reify abstract concepts (such as force), to make them an object the student can manipulate and observe. One of the best examples for such a program is the *Alternate Reality Kit* (ARK, [31]) where students can more or less freely change physical laws and observe the effects on these changes in an animated simulation. ARK does, however, not provide the student with additional tools such as data recording and data analysis tools.

Conclusion. I introduced a computer learning environment that elicits hypothesis-driven learning within the context of a scientific discovery task. This environment is simple yet challenging enough to study certain processes which are central to inductive learning. Clearly, there are many aspects of discovery learning that cannot be observed in REFRACT, in particular the influence of prior knowledge and the development of causal theories. However, I consider REFRACT to be a good starting point for developing computerized learning environments that take into account not only the domain knowledge to be taught, but also psychological insights into processes of learning and discovery.

Acknowledgements

I want to thank Hans Spada, Univ. of Freiburg, for the initial ideas that led to the development of REFRACT and Kalyani Raghavan as well as Robert Glaser, Learning Research and Development Center, for their support during the time the program was designed and implemented.

References

1. Anderson, J. R., Boyle, F. C., Corbett, A. T., & Lewis, M. W.: Cognitive modeling and intelligent tutoring. Artificial Intelligence, 42, 7-49 (1990).
2. Brown, J.S.: Idea-amplifiers: New kinds of electronic learning. Educational Horizons, 63,108-112 (1985)
3. Brown, J.S., Collins, A., & Harris, G.: Artificial intelligence and learning strategies. In: Learning strategies (H. O'Neill, ed.). New York: Academic Press 1978
4. Bruner, J.S.: The act of discovery. Harvard Educational Review 31, 21-32 (1961)
5. Clement, J.: Students' preconceptions in elementary mechanics. American Journal of Physics 50, 66-71 (1982)
6. Collins, A.: Processes in acquiring knowledge. In: Schooling and the acquisition of knowledge (R.C. Anderson, RJ. Shapiro, & WE. Montague, eds.). Hillsdale, NJ: Erlbaum 1977
7. diSessa, A.A.: Unlearning Aristotelian physics: A study of knowledge-based learning. Cognitive Science 6, 37-75 (1982)
8. Holland, J.,, K., Nisbett, R.E., & Thagard, P.: Induction: processes of inference, learning, and discovery. Cambridge, MA: MIT Press 1987
9. Klahr, D., & Dunbar, K.: Dual space search during scientific reasoning. Cognitive Science 12,1-45 (1988)
10. Kuhn, D., Amsel, E., & O'Loughlin, M.: The development of scientific thinking skills. San Diego: Academic Press 1988
11. Kuhn, D., & Phelps, E.: The developmenl of problem solving strategies. In: Advances in child development and behavior (H.W. Reese, ed.). New York: Academic Press 1982
12. Langley, P.: Descriptive discovery processes: experiments in Baconian science. Carnegie Mellon University, Dept. of Computer Science, Doctoral Dissenation 1979
13. Langley, P., Simon, H.A., Bradshaw, G.L., & Zytkow, J.M.: Scientific discovery. Computational explorations of the creative process. Cambridge, MA: MIT Press 1987
14. McDermott, L.C.: Research on conceptual understanding in mechanics. Physics Today 37, 24-32 (1984)
15. Michalski, R., Carbonell, J., & Mitchell, T.M.(Eds.): Machine Learning. An artificial intelligence approach. Palo Alto, CA: Tioga Press 1983
16. Michalski, R., Carbonell, J., & Mitchell, T.M.(Eds.): Machine Learning. An artificial intelligence approach. Volume 2. Los Altos, CA: Kaufmann 1986
17. Mynatt, CR., Doherty, M.E., & Tweney, R.D.: Confirmation bias in a simulated research environment An experimental study of scientific inference. Quarterly Journal of Experimental Psychology 29, 85-95 (1977)

18. Palinscar, A.S., & Brown, A.L.: Reciprocal teaching of comprehension-fostering and comprehension-monitoring. Cognition and Instruction 1,117-175 (1984)
19. Plötzner, R.: Analysis-based induction of elementary functional relationships. Univ. of Freiburg, Dept. of Psychology, Technical Report, April 1990.
20. Pltzner, R., Spada, H., Stumpf, M., & Opwis, K.: Learning qualitative and quantitative reasoning in a microworld for elastic impacts. Univ. of Freiburg, Dept of Psychology, Technical Report, January 1990.
21. Papert, S.: Mindstorms: Children, computers, and powerful ideas. New York. Basic Books 1980
22. Raghavan, K., Schultz, J., Glaser, R., & Schauble, L.: A computer coach for inquiry skills. Univ. of Pittsburgh, Learning Research and Development Center, Technical Report, 1990
23. Reimann, P.: Die Funktion der Hypothese im Lernprozess. In: Handbuch far Schule und Unterricht, Band 8.1 (W. Twellmann, Ed.). Dusseldorf, FRG: Schwann 1985
24. Reimann, P.: Individual differences in hypothesis-guided discovery learning. Observations and a protocol analysis. University of Freiburg, Dept. of Psychology, Technical Repon (1988)
25. Reimann, P.: Problem solving models of scientific discovery learning processes. Frankfurt/M.: Peter Lang Verlag, 1990.
26. Schult, T.: Fallbasierte tutorielle Unterstutzung von Erklarungsprozessen. Univ. of Freiburg, Dept. of Psychology, Technical Report, August 1990.
27. Shultz, RR., & Kestenbaum, N.R.: Causal reasoning in children. Annals of Child Development 2, 159-249 (1985)
28. Shute, V., Glaser, R., & Raghavan, K.: Inference and discovery in an exploratory environment. In: Learning and individual differences (P.L. Ackerman, RJ. Sternberg, & R. Glaser, eds.). San Francisco: Freeman 1987
29. Siegler, R.S., & Liebert, R.M.: Acquisition of formal scientific reasoning by 10- and 13-years-olds: Designing a &ctorial experiment. Developmental Psychology 11, 401-402 (1975~
30. Simon, H.A.: Models of discovery and other topics in the methods of science. Dordrecht: Reidel 1977
31. Smith, R.B.: The Alternative Reality Kit. An animated environment for creating interactive simulations. IEEE Computer Society Workshop on Visual Languages, Dallas, TX, June 1986
32. Spada, H., Reimann, P., & Hausler, B.: Hypothesenerarbeitung und Wissensaufbau beim Schuler. In Kognitive Prozesse und Unterricht. Jahrbuch für Empirische Erziehungswissenschaft(L. Kotter, & H. Mandl, eds.). Dusseldorf: Schwann 1983
33. Stumpf, M., BranskaL S., Herderich, C., Newen, A., Opwis, K., Plotzner, R., Schult, T., & Spada, H.: The graphical user interface of DiBi, a microworld for collision phenomena University of Freiburg, Dept of Psychology, Technical Report, 1988
34. Suchman, J.R.: Inquiry training. Merril-Palmer Quarterly 7,147-169 (1961)
35. Tweney, R.D., Doherty, ME., & Mynatt, CR. (Eds.): On scientific thinking. New York: Columbia University Press 1981
36. Wason, P.C.: On the failure to eliminate hypotheses in a conceptual task. Quarterly Journal of Experimental Psychology 12,129-140 (1960)
37. Wason, P.C.: Reasoning about a rule. Quarterly Journal of Psychology 20, 273-281 (1968)
38. Wason, P.C., & Johnson-Laird, P.N.: Psychology of reasoning: structure and content. Cambridge, MA: Harvard University Press 1972
39. White, B.Y.: Computer games to help physics students understand Newton's law of motion. Cognition and Instruction 1, 69-108 (1984)
40. White, B.Y., & Frederiksen, J.: Intelligent tutoring systems based upon qualitative model evolution. In: Proceedings of the Fifth National Conference on Artificial Intelligence 1986
41. White, B.Y., & Horwitz, P.: Thinker Tools: Enabling children to understand physical laws. Cambridge, MA, BBN Laboratories, Technical Report BBN No. 6470 1987

3. The Interaction Learner/Learning Environment

An Analysis of Cooperation and Conflict in Students' Collaborative Explanations for Phenomena in Mechanics

Michael J. Baker

Centre National de la Recherche Scientifique, Laboratoire IRPEACS, 93 chemin des Mouilles, BP167, F-69131 Lyon-Ecully Cedex, France

Telephone : (33) 72.29.30.12. Fax : (33) 78.33.33.70. Email : MJBAKER@FRMOP11.bitnet

Abstract: In this paper we present a method for analysing students' collaborative explanations for phenomena observed in mechanics experiments. Analyses in terms of the epistemology of the domain and the socio-cognitive processes which operate upon it are presented as a basis for analysing the influence of dialogue processes on explanation generation. The dialogue processes considered are those of maintaining a shared focus, and generating utterances in the context of high-level mutual interaction goals and complex propositional attitudes of speakers. We concentrate on analysing explanation generation in terms of two complex propositional attitudes - *cooperation* and *conflict*. Definitions of these attitudes enable us to identify a number of strategies which students use in 'removing' conflict, and to assess the extent to which cooperation and conflict situations facilitate changes in their beliefs. This analysis is intended to contribute to a computer program for providing automatic guidance which facilitates explanation generation in the framework of collaborative computer-based learning in physics.

Keywords: explanation, physics, dialogue, collaborative learning, cooperation, conflict

1. Introduction

Studying the micro-structure of the way in which students collaborate in constructing explanations is interesting from a number of points of view. In current educational practice, school students often work in pairs, particularly during experimental work in science subjects. However, teachers commonly have little detailed knowledge of what goes on during such sessions. They are usually surprised to find out, and experience in analysing such interactions with teachers can often have an important effect on their teaching. For didactic researchers, looking at pair work provides an opportunity to analyse the epistemological content of students' explanations [47]. In cognitive and social psychology itself, research in collaborative learning

NATO ASI Series, Vol. F 86
Intelligent Learning Environments
and Knowledge Acquisition in Physics
Edited by A. Tiberghien and H. Mandl

situations has been approached from the point of view of the rôle of "socio-cognitive conflict" [18] in cognitive development, and in terms of Vygosky's [51] theories of the rôle of language in learning, where adult guidance or the "scaffolding" provided by collaboration with more capable peers is viewed as an important factor in problem-solving. Given existing educational practice, this research has been applied to a new generation of computer-based learning environments which aim to facilitate collaborative work [28,9,41].

Our own research goals are defined in terms of a specific aspect of collaborative work : we aim to analyse dialogues between pairs of students in order to contribute to the design of a computational guidance system which can facilitate the development of explanations in the context of a computer-based collaborative learning environment in physics. Given learning hypotheses in existing research on the respective rôles of *cooperation* and *conflict* in cognitive development, our research concentrates on analysing the dialogue strategies and structures which learners use in these two situations, and how these strategies relate to the way in which collaborative explanations develop. In order to analyse such dialogue strategies, we need clear definitions of cooperation and conflict, which we provide in terms of *complex mental attitudes*. The principal attitudes of speakers upon which we base our definitions are those of *beliefs* and *goals*. For the purposes of our analysis, a belief is viewed as the attitude of an individual towards a proposition, which may express aspects of the domain - for example, (believes Hélène (red ball_1)) - or other attitudes - for example, (believes Hélène (believes Laurent (red ball_1))). Belief is defined operationally by its *dialogue function* : a belief in some proposition implies that the individual will be prepared to assert the proposition in dialogue and to *argue* for its acceptance. A goal is an attitude towards a proposition which describes a state of affairs, where the individual has a certain willingness to expend resources in achieving it. We also use the notion of *awareness* on the part of an individual towards attitudes, and the notions of *mutuality*, i.e. mutual belief and mutual goals. The definitions of cooperation and conflict in terms of these primitives are discussed in detail later in the chapter. As a preliminary we remark that cooperation is defined in terms of the performance of dialogue utterances which are viewed as contributing towards the achievement of a state of affairs which is a mutual goal. Conflict is defined in terms of awareness of belief in some proposition p, belief that the other speaker believes not-p, or the other speaker does not believe p, or some other proposition is believed to be inconsistent with p. These definitions enable us to describe how utterances which contribute to collaborative explanations relate to cooperative goals, and how successions of utterances transform mental attitudes towards a situation where the conflict situation no longer obtains.

At present we cannot claim any "cognitive status" to our analyses : we rather present research towards developing a new exploratory analysis method, which will be systematically applied to dialogue transcripts in further research. We attempt to apply the method not to dialogues in general, but rather to didactic dialogues between students of a specific age, with a specific problem to resolve. The research is being carried out in collaboration with other researchers[1], within the framework of a wider project which aims to study students' conceptions of energy. The problem chosen is one where pairs of students are asked to *interpret* phenomena observed in mechanics experiments, in terms of a specific concept in physics - that of *energy*. In order to perform this interpretation task, they must generate *explanations*. We now describe our theoretical approach to explanation.

There is a large research literature on explanation in cognitive science, which we can not review here. In order to describe our approach to explanation we identify :

(1) criteria for characterising research goals occurring in the literature ;

(2) recurrent theoretical terms used in analysing explanation ;

(3) a number of different theoretical approaches to explanation analysis.

(1) The two criteria which we use for characterising research goals are :

(i) the nature of the entity which constitutes 'the explanation' ;

(ii) the nature of the entity to be explained.

For example, Schank [42] views 'the explanation' as the individual understanding which results from the process of "intentional reminding", initiated by "expectation failure". The kinds of entities to be explained are physical, social and psychological events and processes. Such psychological preoccupations must be distinguished from research on the linguistic or communicative structure of 'the explanation' as a discourse [27].

(2) The following are recurrent theoretical terms in explanation research :

(2-1) explanation as *understanding* ;

(2-2) *difference* and/or *symmetry* between explanation actors (knowledge, belief, goals, points of view, rights or rôles in the interaction) ;

(2-3) *coherence* of the explanation, at multiple conceptual levels, both 'internal' (an agent's view) and 'external' (coherence with another agent in explanation, with authorities, with the teacher, ...) ;

(2-4) *translation* between representational levels, from a description of the entity to be explained at one level to a coherent view at another level ;

(2-5) explanation as an *emergent* property of interaction.

(3) The theoretical approaches used analysing explanations include *epistemological* (the nature and structure of knowledge in specific domains) *socio-cognitive* (the socio-psychological processes which operate on domain knowledge to generate explanations) and *linguistic* (speech act, discourse structure) approaches.

The specific explanation problem which we consider incorporates these points as follows : (1) 'the explanation' is viewed as the *joint understanding* (2-1) which *emerges* (2-5) from their interaction. "Emergence" means here that participants' explanations (*re* individual mutual and individual beliefs concerning interpretation of physical phenomena) may only emerge during the

---

[1] Prof. A. Tiberghien and F. Langlois, CNRS-IRPEACS.

course of the dialogue. This joint goal is explicit in the specific "social contract" of the interaction.

(2) The speakers must establish the internal and external coherence of their explanations (2-3) in terms of the concept of energy, and establish their relevance to the phenomena to be explained in terms of *intertranslatability* (2-4) with descriptions at an 'objects and events' level. They have *symmetrical* rights and high-level goals in the interaction (2-2), as distinct from a didactic interaction between teacher and student, but there may be *differences* in lower-level specific goals, and in knowledge and beliefs. This may lead to *conflict*. The emphasis of our analysis is therefore on these two interactional situations - COOPERATION and CONFLICT - and their rôle in explanation generation.

(3) We present analyses in terms of three theoretical approaches, which are closely linked : epistemological, socio-cognitive and dialogic. The epistemological analysis assumes a distinction between "everyday terms"[2] and "physics terms". For example, "big" is an everyday term, and "mass" a physics term. " Clearly, this distinction is contentious, and cannot be completely maintained, since terms such as "weight" may be used in either an everday or a physics sense. What it means for a term to be part of physics knowledge is determined by scientific communities (essentially, a Kuhnian view). We adopt a set of epistemological "levels" of analysis, where the syntax and semantics of physics is distinguished (see Tiberghien, this volume). The epistemological analysis essentially defines a *domain knowledge representation scheme*, upon which the *socio-cognitive level* analysis is based. Here we analyse the way in which students jointly intertranslate between different epistemological levels, in terms of schema theory, which we represent in terms of "socially-distributed production rules" [41]. The *dialogic analysis* in turn is based on the cognitive analysis : the aim is to describe how the cognitive processes of explanation generation of each student are influenced by their interaction, and how the joint explanation which emerges from the interaction develops. For example, students' explanation processes are influenced by the necessity to maintain a joint *dialogue focus* [30], by the requirements of managing *turn-taking* [43], and by the general requirements of cooperation [29], such as *relevance* of their utterances to joint goals [45]. Perhaps the most important influence on their explanation generation processes is the additional cognitive load imposed by the necessity to model the attitudes of the other speaker in the dialogue, in order to satisfy these other requirements.

To date there has been relatively little research on the influence of dialogue processes on explanation generation, nothwithstanding Ericsson and Simon's [22] recognition of the need for a model for language generation in studying verbal reports in protocol analysis. Exceptions are the work of Roschelle and his collaborators [9,41] on "transaction analysis" of collaborative learning in a physics microworld, and the research of Barnard and Erkens [10] on modelling the rôle of language production in problem solving. Roschelle *et al* have attempted to use

[2] Later we adopt a more technical definition of "everyday terms" as an "objects and events" representational level.

analyses of turn-taking strategies used by students collaborating in the use of a physics learning environment (the "envisioning machine") to identify when students are really cooperating and constucting a "joint problem space". In our previous work [5,6] we described how turn-taking mechanisms relate to conflict and cooperation. In this chapter we concentrate principally on defining these complex attitudes and the way in which they influence the generation of utterances.

2. Empirical work

Dialogues were recorded and transcribed from children (n = 6) aged 17-18 years, working in pairs. The students were volunteers, working after school hours, with no precise time limit imposed on their activity. Each session turned out to be around 1 and a half hours long. Two observers were present (including their physics teacher), who made notes which were later used for clarification in dialogue analysis. The observers could intervene if the students appeared to be 'stuck'. In our subsequent analysis we consider one specific dialogue ("Hélène and Laurent"). The following problem was given in written form, introduced by the teacher :

> "We want to compare one of the properties allowing a distinction to be made between the substances of which the balls are made. For this, we will study their respective behaviours when they interact with the ground. We will interpret the phenomenon in terms of energy."[3]

The students were given a set of balls of different sizes and substances - rubber, steel, etc. - and also had other materials at their disposal - rulers, weighing scales, wooden planks. They had the possibility to ask for anything else they wanted. After a short introduction by the teacher to the objectives of the experiment, the students discussed together, with few interventions from the teacher.

The objective of the question posed to the students was that they should arrive at an idea of a property of the substance of the balls as the factor which explained their rebound properties, in terms of the "coefficient of restitution" :

vr1 = e.vr2
vr1 = v1 - v'1
vr2 = vé - v'2
v1, v2 velocities before impact
v'1, v'2 velocities after the impact

[3] All problem statements and dialogues appearing in this paper have been translated from the original French by the author. Originals are available on request.

$0<e<1$

$e=0$ perfectly inelastic impact

$e = 1$ perfectly elastic impact

A 'correct' explanation for the rebound of the balls would be something like the following :

Explanation for the rebound of a free-falling ball in terms of Energy

At time the ball is released, t_0, system BALL/EARTH has POTENTIAL ENERGY $= E_p = mgh$.

As the ball falls there is a DISPLACEMENT of the FORCE (mg) through a distance h and there is WORK, which transforms the E_p of the system BALL/EARTH to KINETIC ENERGY = $1/2mv^2$. This energy is transferred to the system BALL

When the ball impacts with the ground the system BALL transfers ENERGY (Mechanical Energy = PE + KE) to the system BALL/EARTH, some of which is converted into HEAT.

The system BALL now has less kinetic energy and so will rebound to a height less than the initial height.

The loss of energy (converted into HEAT) from t_0 to the final time at the highest rebound of the ball, is the difference between the PE of system BALL/EARTH at t_0 (= mgh) and the PE at the highest point after the rebound ($mgh1$; $h1 < h$).

KE is 0 in both of these cases.

In the explanation, physics terms are marked in CAPITALS, to distinguish their use in a physics sense from an 'everyday' sense - a very important point when considering students' conceptions. From students of this age we would not of course expect such knowledge ; but we would expect their explanations to involve the concepts of energy transfer and conversion, work, force and various forms of energy (potential, kinetic, mechanical).

In general, the dialogues follow the pattern of a student proposing an experiment - for example, dropping two balls and measuring the height of the rebound - then discussion on whether the right result had been found, followed by a period during which they discussed their explanations of the results, trying to relate them to concepts of energy with which they were familiar.

3. Analysis of cooperation

We adopt three approaches to analysing the rôle of cooperation in explanation generation :

(i) *epistemological* - in terms of the concepts and methods of physics, in comparison with concepts related to everyday experience ;

(ii) *socio-cognitive* - in terms of individuals' contributions to joint explanations with respect to domain knowledge;
(iii) *dialogic* - in terms of the processes of generation and understanding utterances in explanation dialogues.
The three approaches are clearly interdependent.

3.1 Epistemological analysis

In terms of this approach, we have analysed students' explanations both in terms of
(i) categories derived from scientific methodology, and
(ii) 'representational levels', identified from the referents used with respect to the experimental situation.

3.1.1 Analysis in terms of scientific methodology

The analysis categories used are derived from the scientific methodology as physicists conceive it - in some sense a *normative* description rather than what physicists actually do. We have represented the stages which occur as a "task analysis" [39] in figure 1. Brief definitions are as follows.

Problem interpretation. This includes interpretation of the original written question, and the repeated references to the problem which the students made throughout the course of their problem solving. It is clear that the way in which a problem is represented influences crucially the way in which the problem can be solved.

Validation. This term is used specifically for the case where an experimental hypothesis may be said to have been generated (whether stated explicitly or not), and this hypothesis is interpreted as confirmed or disconfirmed from an experimental result. There are therefore four subtasks to the validation task :

(1) <u>hypothesis</u> and <u>prediction</u> formation ;

(2) proposal and performance of and <u>experiment</u> to test the hypothesis ;

(3) interpretation of the experimental <u>result</u> ;

(4) the result is <u>interpreted</u> as a <u>confirmation</u> or disconfirmation of the hypothesis.

These stages occur rarely in their entirety in the dialogue considered. In the case of one or more steps being absent, although *no validation is said to have taken place,* we may nevertheless use some of the individual subtasks as analytical categories. *Interpretation of the result* is the activity of determining the result and its relationship with prior hypotheses (if any).

As such it is distinguished from *explanation*, although it is clear that there are interrelationships between the two, as follows :

- the hypothesis derived will be influenced by interpretation of the previous results, and by its explanation in terms of energy (when one is given) ;
- previous hypotheses strongly influence perception of the result, in the case where no precise measuring instrument is available;
- the students are rarely clear about any of these stages - we rarely have the situation where they say explicitly "my hypothesis is ...", "the experiment will test this hypothesis and so on.

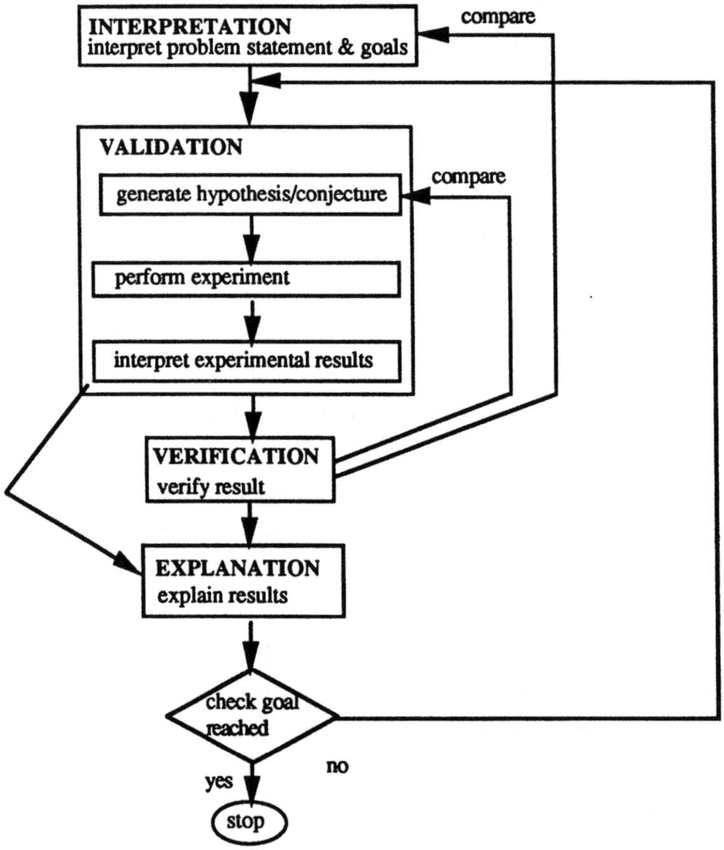

Fig. 1. Task analysis for physics explanation dialogues in terms of elements of scientific method

The following is an extract which illustrates some of these analysis categories.

Example of validation

H: the small one rebounded a bit more!

(result (Hélène) = "rebound different for balls of same materials")

L: well, its not very clear, eh? wait, wait ....

L: the higher one, perhaps?

H: yes but that's not .... the small rebounded ...

L: the small one rebounded more! Now ......

H: yes, but it's, it's the same in fact, they're the same substances which they .... so that shows well, that when the same substance, the volume ... the volume acts no..... the volume,

(interpretation (Hélène) = "when the substance of the balls is the same, their difference in VOLUME is the cause of their different rebound")

or rather the mass

(interpretation (Hélène) = "when the substance of the balls is the same, their difference in MASS is the cause of their different rebound")

Verification. This is the students' activity of "checking" the acceptability of an experimental result with respect to their existing knowledge. This knowledge can be of various kinds [33,34], including references to :

- knowledge of physics ;
- the relationship between physics knowledge and experience ;
- mathematical knowledge (such as the sign of a value obtained from calculation) ;
- the 'didactical contract' (eg using all information in the question; "the teacher said so previously"), and
- the 'social contract' between the students.

This last category is particularly interesting from the point of view of our research objectives in the modelling the influence of interaction between students. For example, they may accept a result as verified if they are agreed upon it. The following is an example of verification which occurs in this dialogue :

Example of verification

"L: I ask myself if we theoretically we shouldn't find a negative value. What we're interested in is the variation in energy. It's positive because at the start it's potential energy. It's greater on arrival."
(verification criterion = mathematical knowledge)

Explanation. We have previously described our point of view on the nature of explanation in this cooperative context. In terms of the epistemology of physics, explanation may be identified with respect to *physics terms used* - in our case, those relating to *energy*. The following is an example of explanation (physics terms marked in **bold** typeface).

Example of explanation

1 H: .... if , if the coefficient of absorption is a function of the **mass** ... the one which has a greater **mass** will have a greater **coefficient of absorption.**

2 L: yes ... but it will fall faster.

3 H: .... so, by ?? it rebounds lower ... but effectively there are the weights which are involved in **potential energy...**

This counts as an explanation both in terms of the physics referents used and in terms of the *position* of the utterances after interpretation of results.

3.1.2 Analysis in terms of representational levels

Analysis in terms of representational levels gives us some means of defining the explanation process itself, in contrast to *interpretation* of results, since utterances forming parts of explanations are expressed in terms of concepts related to *energy*, whereas *interpretations* and descriptions of results are usually expressed in terms of "objects and events" (for example "heavy", "weight", rather than "mass"). Explanation in this context can thus also be viewed in terms of *translation* between descriptions in everyday terms to descriptions in terms of physics. The explanation process is thus closely linked to that of development of the *problem representation* in problem solving tasks. Chi, Feltovich and Glaser [14] have described four "representational levels" in the creation of problem representations in mechanics :

Level 1 : physical objects, properties and problem statement referents ;

Level 2 : associative network structures for representations at level 1 ;

Level 3 : physics concepts and idealised objects ;

Level 4 : the syntax and semantics of physics equations.

In the following example we can analyse the way in which students cooperate in the process of 'translation' between levels (referents are in **bold**).

Example : representational levels in collaborative explanation dialogues

1."H: and uh, and we can say that .... wait, what are these...

2.L: ok, in fact we know that there is **conservation of quantities [level 3]** during movement...

3.H: look! this one's a lot **softer [level 1]** than that one, but that one is softer than the **steel** [level 1] one.

4.L: ah, ah yes.

5.H: so, the **steel** [level 1] rebounds more than the **rubber** [level 1] , uh, **yellow** [level 1] one there, and the **black** [level 1] one didn't rebound at all.

6.L: its not a matter of ..... of **density** [level 3] , I think, all the same, it's not the density.

7.H: good, the difference between the substance which the ball's made of!

8.L: yes, its the subst ..... yes!

9.H: ah! so, in terms of **energy** [level 3] , we can see that ...

10.L: right, hang on, support !

(throws ball)

11.H: hang on, because there's the .... if we release at a certain height, there's **potential energy** involved .....[level 3]

12.L: yes.

13.H: with the, the **weights** [level 1] , with the **mass** [level 3] .

14.L: could we say that there is conservation? ....no, because there are fr[ictions] ..... ok, right, if we ignore, we said we'd ignore air friction, right, in fact ....."

It should be noted that since a given term such as "weight" may be used at either level 1 or 3 (line 13 above), its interpretation is *context dependent*. Such translations work in both directions. At line 2 L describes the result in terms of a physics concept ; however, H is presently thinking of the result at level 1. At 9 H attempts to translate these descriptions into the physics term "energy", after both L and H have described the situation in terms of the properties of the balls. At 11, H is also contributes to the explanation at the same level. In 13 H makes a self-correction, in translating the everyday term "weight" into "mass", presumably because she knows that the former would not be consistent with a physics level description (it does not appear in the energy equations, but "mass" does). In accordance with well known results in physics education (see [19]) we observed that students generally 'dwell' on the objects and events level - the colour of the balls, their size, the floor - and make repeated attempts to relate these descriptions both to energy equations and the descriptions in the original problem statement.

3.2 Socio-cognitive analysis

The epistemological analysis provides the general domain content of students' explanations, upon which we may base a model for socio-cognitive processes in explanation. For *individual* students, these explanation generation processes have been described in the literature in terms of bottom-up and top-down mechanisms in *schema matching* [14]. We have attempted to apply this production rule analysis method to the complete dialogue between Laurent and Hélène (approximately 60 rules have been identified), with the objective of exploring how students cooperate in construction of explanations, at the level of the domain concepts used, derived from the epistemological analysis. In many cases it proved impossible to analyse such rules without recourse to both students in interaction. Roschelle & Behrend [41] term rules analysed in this way "socially-distributed productions". They argue that instead of viewing production rules as somehow "inside the head" of each individual student, a more fruitful approach is to analyse the way in which collaborative problem solving "takes place *between* the partners". In addition to the "collaborative completion" which they identify, where in effect a compound sentence is distributed across speaker turns, we identify various forms of *collaborative elaboration*. From these analyses we have identified the following phenomena :

(1) collaborative completion of rule right-hand side (rhs) by the other student ;

(2) collaborative elaboration of the rhs of rules by the other student ;

(3) collaborative elaboration of the left-hand side (lhs) of rules.

The following are some examples.

Collaborative completion of rule right-hand side

"L: and kinetic, and in fact, the kinetic energy... it...	**IF kinetic energy is involved**
H: and well, it, it is nil at the start	**THEN kinetic energy is nil at the moment it is released**

Collaborative elaboration of the rule right-hand side

"L: so, what can we say if there's an inelastic impact?	**IF there is an inelastic impact**
H: um, well, that the energy, ... all the energy ..;	**THEN the energy ...**
L: well, the kinetic energy, is theoretically nil.	**the kinetic energy is nil ...**
H: it's nil on arrival, in fact"	**on arrival at the impact**

"L: what is it in ... in the characteristic of this	**IF one ball rebounds, and the other does not**
ball which is going to ...going to tell us that	**THEN characteristic which explains rebound is ...**
that one absorbs and that one rebounds?	
H: well, it's what we called the rigidity"[4]	**the rigidity**

[4] Note that the students invented this term themselves - it was not part of taught knowledge.

Collaborative elaboration of the lhs

"L:ok, there's a total loss of velocity,	**IF there is total loss of velocity**
which means that ...	**and**
H: wait, wait, ... with their interaction,	
with the ground.	**there is interaction with ground**
L: so the ground absorbs eh...	**THEN the ground absorbs ...**
H: yes	
L: absorbs the impact!	**the impact**

In this example, we also see collaborative elaboration of the rhs.

In all of these cases it is not easy to isolate *why* this collaboration takes place : is a right-hand side completed by another because the first student cannot complete it, or has the second student taken an opportunity (perhaps a hesitation on the part of the first) to *interrupt* and supply something relevant, which was not in fact what the first student would have said? A number of rhs of rules can be interpreted as calls to "action schemata" [14], being procedures for solving equations attached to problem schemata :

"H: so, wait, it leaves with a velocity of nil at 1metre height, ...
L: wait, the energy, the mechanical energy, uh, the mechanical energy is the potential energy plus the kinetic energy"

[Hélène]

IF **there is velocity and height involved**
THEN **(...)**

[Laurent]

THEN **[apply action schema : calculate the mechanical energy by adding the potential energy to kinetic energy]**

Finally, we analysed one case where the student could clearly see that a particular schema applied, but did not articulate the conditions which led her to believe this to be the case :

"H: I'm sure there's a proportionality somewhere..."

IF **(...)**
THEN **there is proportionality involved**

The analysis which we have described forms the basis for a cooperative production rule model for explanation in this restricted domain. We now analyse students'propositional attitudes towards their representations of the domain in order to model the processes by which they generate utterances in a cooperative dialogue context.

3.3 Dialogue level analysis

At this level we aim to analyse how the specific utterances which students generate to express their explanations relate to the set of implicit assumptions of cooperative dialogues. One such feature is the managing of turn-taking [43]. Since this has been extensively analysed by Roschelle and Behrend [41] in similar dialogues, we concentrate here on analysis of individual utterances as *speech acts* [44], and attempt to describe the way in which they are generated from mental states such as propositional attitudes [24], in order to satisfy cooperative goals in the interaction. We also analyse *conflict* situations in terms of complex propositional attitudes, and specific kinds of utterances used in *argument*.

3.3.1 Analysis of cooperation

The general set of assumptions of speakers in cooperative interactions has been analysed by Grice [29] in terms of a set of "conversational maxims" for sincerity, informativeness, and so on. In our analysis we assume Perrault's [37] formalisation of a number of these maxims. We also assume the general "deliberative agent" architecture [26] underlying production system models, a set of propositional attitudes derived from general theories of rational agency and a set of utterance categories derived from computational speech act theory [17]. The analysis method consists of assigning the utterance categories, together with their propositional content to the dialogue, and showing how they may be generated from propositional attitudes together with representations of the high-level goals of the interaction. In the case of analysing conflict, we assign corresponding complex attitudes to the students at given points in the dialogue. A production system is currently being programmed which can generate some utterance types from representations of initial domain knowledge and attitudes of students. The agent attitudes used include *beliefs*, *intentions*, *goals*, *committments*, *preferences* and *wants*. There are of course many profound theoretical issues which are currently being addressed in agent theory, concerning the nature of belief and intention, for example. Rather than performing *a priori* analyses, our research strategy is to take existing formalisations and to successively refine them with respect to analyses of real dialogue protocols. Here we can only give a short illustration of the model currently being developed. The 'content' of propositions referred to in the analysis is derived from the previous epistemological analysis.

An important starting point for the analysis is the set of mutual goals which constitute the "social contract" between the students :

L and H have the mutual goal that they will mutually believe that some set of propositions {p1, ... pn} explain a given set of results {r1, ... rn} (G1) for a given set of experiments {expt_1,....expt_n}, that they will agree upon these results (G2) and that the propositions be instances of some subconcept of energy (G3) :

Mutual goals

G1 : (mgoal L H (mbelief L H (explains {p1, ... pn}
(rebound_behaviour (result {r1, ... rn} {expt_1,....expt_n}))))).

G2 : (mgoal L H (mbelief L H ({r1, ... rn} {expt_1,....expt_n}))).

G3 : (mgoal L H (mbelief L H (instance {p1, ... pn} (subconcept energy)))).

The definition of mutual goals ("mgoal") is dervived from the work of Galliers [25] and Elsom-Cook [21] :

R1 : (mgoal S1 S2 g) --> (goal S1 g) & (believes S1 (goal S2 g)) &
(goal S2 g) & (believes S2 (goal S1 g)).

We assume that the well-known infinite regress of mutual beliefs - (believes S1 (believes S2 (believes S1 .... - is stopped at the first level of embedding :

R2 : (mbelief S1 S2 b) --> (believes S1 b) & (believes S1 (believes S2 b)) &
(believes S2 b) & (believes S2 (believes S1 b)).

Let us now look at an example utterance pair :

Utterances	Analysis
"L: it rebounds all the same a bit lower ...	(INFORM L H p1)
	p1 : (result (lower_than (rebound ball_1) (rebound ball_2)))
H: yes,	(EXPRESS H L (believes H p1))

but with respect to the masses
you can see that the steel one is heavier"

Definitions of the categories INFORM and EXPRESS are derived from computational pragmatics : to INFORM some proposition is to ASSERT it with the intention that the proposition be believed by the hearer [37]. To ASSERT a proposition is to intend that the hearer believes that the speaker believes the proposition. EXPRESS is a category of utterances where the speaker communicates their *attitude* to some proposition, in this case, *belief,* [49]. We need to show how this utterance analysis can be derived from the goals, and from *reaction* to the other speaker's utterances. Assume that a focus mechanism selects p1 as the most relevant proposition, both in terms of other propositions in semantic memory and in terms of the high level goals [30]. From G3, R1 and R2, and a rule which states that the components into which a goal may be analysed are also goals

R3 : (goal S1 x) --> (goal S1 (subcomponents x)).

we can derive

(goal L (believes H p1)) &
(goal L (believes H (believes L p1)))

Many other goals may be derived - such as that L should believe p1 - which are already satisfied. Now, if there are no other goals to which L is *committed* which L believes conflict with these goals, L may become committed[5] to them. Assume therefore that L is committed to these goals. If an agent is committed to a goal, then the agent *wants* to perform a dialogue action which it is believed will contribute to its achievement. Now, as a definition of INFORM we have :

R4 : (wants A1 (believes A2 x)) & (wants A1 (believes A2 (believes A1 x))) &
(believes A1 x)
--> (wants A1 (INFORM A1 A2 x).

So, L wants to INFORM H of p1. Choice amongst wanted actions is made according to a preference mechanism will evaluates the extent to which an action is likely to be successful in achieving the wanted goal. This involves reasoning about the propositional attitudes of the other speaker. We have described how a set of rules may be developed which relate attitudes to dialogue utterance types. We also require rules for updating the speaker's beliefs about the

[5] We adopt a definition of commitment in terms of Cohen and Levesque's [16] definition of intention

beliefs of the other, subsequent to performing an action, and for hearer's to infer attitudes from utterances of the other. Now assume that from the communicative context, the syntactic form of the utterance and its intonation contour [38], H is able to recognise it as a declarative sentence. From a combination of this and interpretation of the utterance in relation to high-level goals, H infers that L wants her to believe p1 and that L believes it (sincerity assumption) :

R5 : (believes S1 (INFORM S2 S1 x)) -->
(believes S1 (want S2 (believes S1 x))) & (believes S1 (believes S2 x)).

Now we can assume that H does not believe p1 to be *inconsistent* with her beliefs, but not that she actually believes it. If she believes that L has INFORMED p1, she believes that L wants her to believe it and that L believes it (the reverse of R5). Since it is her turn, she first responds, then makes another declaration. So we have :

R6 : (believes S1 (wants S2 (attitude? S1 x))) -->
(wants (EXPRESS S1 S2 (attitude? S1 x))).

So L expresses her attitude (believes L p1) in the word "yes". We need to look at the wider cooperative context in order to determine whether the "yes" expresses the attitude of belief or not, since L could be simply 'taking her turn' in the dialogue. In this case, her response concerns the experimental result rather than its explanation ; since there is no evidence later in the dialogue to suggest that she does not believe this interpretation of the result, we assume this interpretation of her utterance. This illustrates one of the profound difficulties of this kind of analysis, and the necessity to consider the wider dialogue context.

We have not described all aspects of the model which we are developing - the focus mechanism, mechanisms for preferring wanted actions to others, what it means for goals to be inconsistent, and so on - but have rather given a short example to illustrate the approach. In the context of explanation, the most important question to be addressed is how and when agents are willing to adopt beliefs stated by others - whether they are believed to be inconsistent with their own or not. Our approach to this is based on existing work by Harman (1986). We also analyse the phenomenon of belief revision in terms of *conflict situations*.

3.3.2 Analysis of conflict

The incidence of inter-individual conflict in problem-solving tasks has been argued to be a necessary condition for cognitive development, under certain specific conditions [18]. It is possible, therefore, that conflict and its resolution plays a constructive rôle in the development of cooperative explanations. Results from the study of "socio-cognitive conflict" on development are, however, inconclusive, and some research suggests that conflict is associated with *less* progress [11]. Our preliminary qualitative analyses suggest that the precise way in which conflict is dealt with may be important in determining the extent to which students' beliefs may change. Research on the formalisation of conflict resolution exists within the framework of "Distributed Artificial Intelligence" [46], where it has been approached from the point of view of specifying *negotiation protocols* for resolving goal conflicts and task allocation problems between distributed agents. Little of this research has addressed the problem of resolving conflicts in *beliefs* or *points of view*, which has rather been approached from the perspective of *argumentation* research. This has been concerned with identifying argument units and their interrelations [1], linguistic markers associated with arguments [20,36], grammars for understanding arguments [15], and with defining formal dialectics systems [8].

In our approach we characterise conflict and its resolution in terms of the model described earlier for cooperation at the dialogic level - in terms of complex propositional attitudes and the speech acts which they generate. We define several different kinds of conflict, which allows detailed analysis of the various ways in which conflict may be 'removed' or 'resolved'. We adopt definitions of conflict based on modifications of the research of Galliers [25] and Elsom-Cook [21] as follows.

Internal conflict is a notion of conflict within the beliefs of a single individual. It depends on the notions of *awareness* and *inconsistency* :

Definition of internal conflict

(believes A1 X1) & (believes A1 X2) & (aware A1 X1) & (aware A1 X2) &
(believes AI (inconsistent X1 X2))
--> (**int_conflict** A1 X1 X2).

[A1,A2 = agents, X1,X2 = propositions]

"Inconsistency" is not necessarily a matter of logical contradiction - it may relate to a more general *coherence model* of belief revision [31]. "Awareness" is modelled as focus of attention (Grosz & Sidner 1986) in working memory, within the general cognitive architecture. We have applied the notion of "spreading activation" [2] to modelling focus in tutorial dialogues [4,6].

External conflict concerns conflict between beliefs of two or more different agents. There are at least two different kinds - where some believed conflicting alternative proposition has been stated (ext_conflict_a"), and where a simple denial has been stated with no reason given ("ext_conflict_d"). Here we differ with Elsom-Cook [21], since these two different kinds need to be distinguished in the dialogues analysed.

<u>Definitions of external conflict</u>

(believes A1 X1) & (aware A1 X1) &
(believes A1 (believes A2 X2) & (aware A1 (believes A2 X2)) &
(believes A1 (inconsistent X1 X2))
--> (**ext_conflict_a** A1 A2 X1 X2).

(believes A1 X1) & (aware A1 X1) &
(believes A1 (believes A2 (not X1)) & (aware A1 (believes A2 (not X1)))
--> (**ext_conflict_d** A1 A2 X1).

These forms of external conflict may also be *mutual*, for example (and similarly for **m_ext_conflict_a**) :

<u>Definition of mutual external conflict</u>

(**ext_conflict_d** A1 A2 X1) & (**ext_conflict_d** A2 A1 X1)
--> (**m_ext_conflict_d** A1 A2 X1).

The process of conflict resolution is one example of *negotiation* [4,6], which we define as :

<u>Definition of negotiation</u> : the set of cognitive processes which generate a sequence of dialogue actions by two speakers, who possess the propositional postures of indifference or conflict with respect to one or more propositions, and one or more of whom possesses the goal of transforming the mutually shared posture to one of cooperation with respect to some proposition.

Note therefore that negotiation is not exclusively concerned with conflict resolution[6], nor with goal conflict [6], but may include the attempt to secure mutual belief from a position of indifference or absence of such. Note also that initial attitudes of speakers do not have to be symmetrical - one could believe conflict to exist and the other not, for example.

Consider now the definition of ext_conflict_a. This complex attitude ceases to be ascribed to an agent if any combination of the conjuncts in the definition is no longer ascribable (there

[6] Sycara [46] restricts negotiation to "...the process of achieving compromise between conflicting goals".

are $2^4 - 1 = 15$ possibilities). For each specific conflict situation we therefore need to ascribe the appropriate conflict attitude, to describe which combination of conjunctive primitive attitudes no longer hold on conflict removal, and to describe the dialogue processes which led to this removal. As a preliminary means to exploring the viability of the analysis method, we have analysed conflict situations in the Hélène/Laurent dialogue. We can only give a few examples here, taken from a consecutive series of 39 exchanges near the beginning of the dialogue (one and a half hour duration). The definitions are presently rather simple, but prove adequate for analysing the phenomena observed here.

In a more complete analysis we would also need to make use of descriptions of dialectic moves in argument - support, counterclaim, etc. For example, in addition to analysing a specific utterance as an ASSERTION, we need to represent the fact that the utterance may be viewed as *supporting*, or *denying* some previous utterance. A method for analysis such phenomena may be derived from the work of Toulmin, Rieke and Janik [48] and Barth and Krabbe [8]. These do not constitute new speech act categories, but are rather *relations* between them [23]. There are different types of relations : speech acts may enter into an *explanatory* relation with the *propositional content* of others, but in the *justification* relation, what is justified is the entire act/proposition itself.

<u>Conflict example 1</u> (p1, ... = propositions)

1 L: it rebounds all the same a bit lower, it's normal, but still ...

2 H: yes, but look, with respect to the masses, look you can see that the steel one is ... heavier

3 L: yes, but it's not a question of mass.

p1 : (explains "mass" (lower (rebound ball_1) (rebound ball_2)))

(ext_conflict_d L H p1)

4 H: well ; and , there's all the same potential energy involved, I'm sorry !

(ext_conflict_d H L p1)

(m_ext_conflict_d L H p1)

5 L: ...... ok, but if you've ...

6 H: if we re...

7 L: if you had a big steel ball, it would rebound ...

8 H: if we release them at the same height, so that there's one which has a greater mass than the other, the one with the bigger mass, would have greater potential energy ...

9 L: yes, but ...

10 H: so, there will be more ...

11 L: do you believe that if, if you have an enormous rubber ball like that, which is a kilogramme, you believe it would rebound a lot ?

12 H: yes, but that's only valid in the case of a perfectly elastic impact.

13 L: right, ...

14 H: in fact, I think ...

15 L: perhaps we'd be better off thinking about that, because, a priori, it's simpler, given that it's an inelastic impact.

The first conflict occurs at 3 and 4, where L denies H's (indirect) claim that the difference in rebound behaviour is explained by the *mass* of the balls. Since L does not propose an alternative explanation, this is analysed as **ext_conflict_d,** which, when recognised by H becomes **m_ext_conflict_d.** From our earlier analysis of conflict (**ext_conflict_d**), minimal attitudes of H and L are as follows :

<u>Hélène</u> : (believes H p1) & (aware H p1) &
(believes H (believes L (not p1)) & (aware H (believes L (not p1)))

<u>Laurent</u> : (believes L (not p1)) & (aware L (not p1)) &
(believes L (believes H p1)) & (aware L (believes H p1))

Our analysis therefore predicts that mutual conflict is transformed to individual conflict when any combination of the conjuncts for either H or L ceases to hold, and that mutual conflict is removed when any combination of conjuncts no longer holds in the complex attitude for each. In this case, it is difficult to see why L disagrees. H produces *support* for her beliefs by stating that potential energy is involved, to which L *concedes*. H then produces a further supporting argument (the greater the mass, the greater the potential energy and rebound), to which L again concedes, but produces a *plausible counter-argument* (counter example of large rubber ball , not rebounding much), to which H does not concede in the case of an inelastic rebound, but concedes in the case of a perfectly elastic rebound. So ; we are now in a situation where we might expect conflict to be removed by L conceding p1 in the case of a perfectly elastic rebound. This does not appear to be the case. The specific conflict resolution which occurs here is as follows :

- since L has produced no direct successful counterargument against H's argumentative support for p1, H *assumes* that L accepts p1 [(believes H (believes L p1))] ;
- as shown later in the dialogue, in fact (believes L (not p1))
- the *focus of attention* shifts to considering the case of elastic and inelastic impacts, so we have [(not (aware L (not p1))) & (not (aware H p1))]
- the *cognitive resources* which H and L are willing to use in resolving this conflict are *expended*, and so, they explicitly signal a focus shift (line 15 : " L: perhaps we'd be better off thinking about that ... ").

In this case, therefore, we observe that conflict may be 'removed' by :

- concentrating on related propositions with respect to which agreement occurs, thus no longer being aware of propositions with respect to which conflict occurred ;
- no longer being aware of conflict by focussing on a new topic introduced in the argument when cognitive resources are expended ;
- believing that the other implicitly agrees by assuming an argument to be convincing in the absence of effective counter-arguments to supports for other claims ;
- believing that the other implicitly agrees by assuming an argument to be convincing in the absence of alternative plausible claims.

Let us now consider a second example.

<u>Conflict example 2</u>

28 H: yes, elastic impact, there's the total energy which is conserved.

29 L: yes ...

30 H: yes, but there's the friction ... of the air !

p2: (explains (friction air)

(result (not (same (rebound ball_1) (rebound ball_2)))))

31 L: oh, I don't think its especially the air friction which invervenes ...

(ext_conflict_d H L p2)

32 H: but yes, otherwise it would rebound to the same height !

(m_ext_conflict_d L H p2)

p3 : (same (rebound ball_1) (rebound ball_2)) [stated earlier in dialogue]

p4: ((((not p2) --> p3)) & (not p3)) -> p2)

33 L: no ! ...

(ext_conflict_d L H p4)

34 H: but yes !

(m_ext_conflict_d L H p4)

35 L: ... it's the loss at the moment of impact ...

p5: (explains loss_at_impact

(result (not (same (rebound ball_1) (rebound ball_2)))))

36 H: that's the same ... that's really a sort of friction also. It's a form of friction, either with the ground or with the air

(ext_conflict_a L H p5)

p6: (equivalent loss_at_impact (friction air))

37 L: yes, but after all, air friction in comparision with friction ... no, no, what you call ground friction is by contrast negligable !

(ext_conflict_a L H p6)

38 H: ah well .... yes

This is a complex sequence. What we have in effect is a succession of conflicts, with only the final one resolved by explicit acknowledgement. Even in this case we cannot assume that H's "yes" implies that they genuinely adopt L's point of view. It divides into two phases : 28-34, where conflicting propositions are simply *denied*, and 35-38 where alternative propositions ("counterclaims") are proposed. Each of the conflicts have conjuncts in their definitions concerning awareness and belief in the conflicting propositions. Both of these are involved in resolution of these conflicts - i.e. the speaker who 'won' the last conflict resolution (L) assumes that H no longer believes the conflicting propositions, since he believes this last argument to be *won*, and neither student is now *aware* of the original propositions with respect to which conflict took place. Therefore, L assumes that winning an argument means that L and H are agreed on some proposition p_x, that p_x is inconsistent with some other proposition p_y, that this belief in inconsistency will be shared by H, and finally that this will lead to H dropping their belief in p_y.

With respect to the possible effects of conflict in the evolution of students' explanations, it is clear that their explanations do not evolve as effectively as they might since : learners have a tendency to 'remove' conflict rather than to 'resolve' it. We summarise the ways in which students remove conflict as follows :

focus shifts

- focussing on related propositions with respect to which there is agreement ;
- focussing on the most recent conflicts ;
- focussing on recent concluded local arguments ;
- (focus shifts may sometimes be occasioned by *exhaustion of cognitive resources* when conflict is not resolved after a certain time).

belief revision assumptions

- believing that a 'won' local argument implies belief revision ;
- believing that an unsupported alternative claim implies acceptance of their own claim ;
- believing that absence of an alternative claim implies acceptance ;

lack of clarity

- students' understanding of their own arguments and beliefs, as well as of those of the other, is unclear, which leads to an unclear understanding of the conflict situation (the propositions with respect to which conflict takes place).

Assumptions concerning belief revision and mechanisms for shifting focus are therefore important elements in explaining students' conflict resolution strategies in this domain.

It is understandable that conflict between students tends to be *removed* rather than *resolved* and that beliefs are not genuinely revised in such conflicts : in the absence of an arbiter, such as a human teacher, there is no direct *legitimisation* of their beliefs, although students do employ various strategies for *validating* and/or *verifying* their results and explanations [33,34], with respect to their own knowledge, and aspects of the general "didactic contract" ("all information in the question is relevant", "the teacher said so", etc.). We can therefore hypothesise a clear role for the use of AI techniques for generating automatic guidance for facilitating students' explanations. We believe that the following kinds of adaptive guidance may be useful with respect to resolving students' conflicts :

- identification of reasonable and unreasonable arguments ;
- identification of propositions with respect to which conflict takes place ;
- clarification of the beliefs of each student ;
- identification of dialogic or argumentative relations between propositions stated ;
- recapitulation and reinforcement of argument conclusions ;
- indication as to when an argument is worth pursuing or not.

What this amounts to is the claim that the generation of effective negotiated explanations between students requires a system for providing guidance or tutoring on *how to argue*. Providing automatic guidance of this kind within the context of cooperative computer-based learning environments needs to be situated within a more general cognitive simulation of explanation and argument in dialogue.

4. Conclusions and further work

In this paper we have described an integrated approach to analysing students' cooperative explanations. The method is intended to take the influence of the dialogue processes on explanation generation into account, and to produce an analysis which can form the basis of a computationally tractable model for providing automatic guidance for cooperative explanation generation. Our present concern has been to define a set of suitable analysis methods, which will be systematically applied to further dialogue protocols in further research. Such an enterprise is interesting in its own right, but a precise analysis may also help us to define reasons why explanations do not develop effectively, and thus a rôle for computer-based education in the context of collaborative computer work.

Our qualitative analysis suggests that cooperation is more constructive than conflict in the evolution of students' explanations. In their search for agreement, students may settle on explanations which neither would have generated alone. In the absence of confirmation and legitimisation, students often do not pursue the most fruitful directions amongst those which have been stated. There is therefore a clear need for guidance in *cooperative* contexts. We also argued that students in fact rarely *resolve* conflicts, since they either focus on subclaims upon which there is agreement, change focus to the most recent conflicts, assume that the conflict is resolved if the most recent sub-argument chain is 'won', or simply give up. Conflict itself is rarely clearly understood, which leads to 'softening' its perception. Students could therefore benefit from guidance in 'sharpening' their understanding of each other's beliefs and claims, and in indicating when a conflict is worth resolving. We therefore propose the following rôle for computer-based guidance.

The learning context considered is *computer-based distance education*, where each student has access to a mechanics learning environment for the problem we desribed. Within the environment, students can communicate their explanations in cooperation and conflict situations to each other. We shall use a deliberately *restricted* interface for communication, which forces students to clearly state the utterance and proposition units of their explanations and arguments. Such a tool has been described for arguments by VanLehn [50]. Forcing students to express themselves in this way should eliminate some of the ambiguity of the conflict situation derived from natural language interaction. Within the system, a computer 'monitor' program will generate guidance for both students by explaining the utterances and beliefs of each to the other, indicating which cooperative explanations are worth pursuing, and which conflicts are worth resolving. It may also need to provide new information in order to do so, using a domain model, the elements of which are provided in our analyses at the epistemological and socio-cognitive levels. These are our ambitious long-term research goals, for which the analyses presented provide good theoretical and empirical bases.

Acknowledgements

Many thanks to Andrée Tiberghien for constructive criticism throughout this research, and to Françoise Langlois for generously making the dialogues available. This research is supported by the Centre National de la Recherche Scientifique.

References

1. Alvarado, S.J., Dyer, M.G. & Flowers, M. : Editorial Comprehension in OpEd through argument units. Proceedings of the Fifth National Conference on Artificial Intelligence, 250-256, Philadelphia USA, 1986.

2. Anderson, J.R. : The Architecture of Cognition. Harvard : Harvard University Press, 1983.
3. Baker, M.J. : A model for tutorial dialogues based on critical argument. Artificial Intelligence and Education, (Eds.) Bierman, D., Breuker, J. & Sandberg, J. Amsterdam : IOS Publishers, 1989a.
4. Baker, M.J. : Negotiated Tutoring: A Model for Interaction in Intelligent Tutoring Systems. PhD thesis in Cognitive Science, (Unpublished). Institute of Educational Technology, Centre for Information Technology in Education : The Open University, Great Britain, 1989b.
5. Baker, M.J. : Arguing with the tutor: a model for tutorial dialogue in uncertain knowledge domains. In Guided-Discovery Tutoring, (ed.) Elsom-Cook, M. , p. 141-170. London : Paul Chapman Publishing, 1991a.
6. Baker, M.J. : Negotiating Goals in Intelligent Tutoring Dialogues. To be published in New Directions in Intelligent Tutoring Systems, (ed.) Costa, E., NATO ASI Series, 1991b, in press.
7. Baker, M.J. : The Influence of Dialogue Processes on Students' Collaborativbe Generation of Explanations for Simple Physical Phenomena. In Proceedings of the International Conference on the Learning Sciences, Evanston Illinois USA, 1991c, in press.
8. Barth, E.M. & Krabbe, E.C.W. : From Axiom to Dialogue : A Philosophical Study of Logics and Argumentation. Berlin : Walter de Gruyter, 1982.
9. Beherend, S. Singer, J. and Roschelle, J. : A Methodology for the Analysis of Collaborative Learning in a Physics Microworld. Proceedings of ITS '88, 48-53, 1988.
10. Barnard, Y. and Erkens, G. : Simulation and Analysis of Problem-Solving and Dialogue Processing within Co-operative Learning. Learning and Instruction : European Research in an International Context. Volume 2:1, Social and Cognitive Aspects of Learning and Instruction, (eds.) Mandl, H., De Corte, E., Bennett, S. N. & Friedrich, H.F., pp. 181-196. London : Pergamon Press, 1990.
11. Blaye, A. : Peer Interaction in Solving a Binary Matrix Problem : Possible Mechanisms Causing Individual Progress. In Learning and Instruction Vol 2,1, (eds.) Mandl, H., De Corte, E., Bennett, N. & Friedrich, H.F., pp. 45-56. Pergamon Press, 1990.
12. Chan, T.W. and Baskin, A.B. : Studying with the Prince : The Computer Companion as a Learning Companion. Proceedings of ITS-88, Montréal, 1988.
13. Chi, M.T.H., Bassok, M., Lewis, M.W., Reimann, P. and Glaser, R. : Self-Explanations: How Students Study and Use Examples in Learning to Solve Problems. Cognitive Science 13 (2), 145-182, 1989.
14. Chi, M.T.H., Feltovich, P.J. and Glaser, R. : Categorization and Representation of Physics Problems by Experts and Novices. Cognitive Science 2, 121-152, 1981.
15. Cohen, R. : Analyzing the Structure of Argumentative Discourse. Computational Linguistics, 13 (1-2), 1-24, 1987.
16. Cohen, P.R. & Levesque, H.J. : Intention is Choice with Committment. Artificial Intelligence 42, 2-3, 213-261, 1990.
17. Cohen, P.R., Morgan, J. and Pollack, M.E. : Intentions in Communication. Cambridge, Mass. : MIT Press, 1990.
18. Doise, W., Mugny, G. & Perret-Clermont, A.N. : Social Interaction and Cognitive Development. European Journal of Social Psychology, 5, 367-83, 1975.
19. Driver, R., Guesne, E. & Tiberghien, A. : Children's Ideas in Science. Milton Keynes : The Open University Press, 1985.
20. Ducrot, O. : Les Echelles Argumentatives. Paris : Minuit, 1980.
21. Elsom-Cook, M. : Analysis of a Tutorial Dialogue. In Elsom-Cook, M. (ed.) Guided Discovery Tutoring, p. 113-131. London : Paul Chapman Publishing, 1990.
22. Ericsson, K.A. & Simon, H.A. : Verbal Reports as Data. Psychological Review 87 (3), 215-251, 1980.
23. Ferrara, A. : Pragmatics. In Handbook of Discourse Analysis, Vol. 2 : Dimensions of Discourse. London : Academic Press, 1985.
24. Fodor, J.A. : Representations. Cambridge MA : MIT Press, 1981.
25. Galliers, J.R. : A Theoretical Framework for Computer Models of Cooperative Dialogue, Acknowledging Multi-Agent Conflict. PhD thesis (unpublished), Human Cognition Research Laboratory, The Open University (GB), 1989.
26. Genesereth, M.R. & Nilsson, N.J. : Logical Foundations of Artificial Intelligence. Los Altos, Calif. : Morgan Kaufmann, 1987.
27. Goguen, J.A., Weiner, J.L. & Linde, C. : Reasoning and Natural Explanation. International Journal of Man-Machine Studies 19, 521-559, 1983.
28. Greif, E. : Computer-supported cooperative work : a book of readings. Morgan Kaufmann, 1988.
29. Grice, P.H. : Logic and Conversation. In Syntax and Semantics, Volume 3 : Speech Acts, (Eds.) Cole, P. & Morgan, J.L. New York : Academic Press, 1975.
30. Grosz, B.J. and Sidner, C. : Attention, Intentions and the Structure of Discourse. Computational Linguistics, 12 (3), 175-204, 1986.
31. Harman, G. : Change in View. Cambridge Mass. : MIT Press, 1986.
32. Klahr, D., Langley, P. & Neches, R. : Production System Models of Learning and Development. MIT Press, 1987.

33. Langlois, F. : Problèmes de Physique en 1°S: Etude de Quelques Critères de Vérification. Mémoire de DEA, dir. A. TIBERGHIEN, Groupe Didactique de Physique, Université Lyon I, 1990.
34. Langlois, F. and Tiberghien, A. : Résolution de Problèmes de Physique par des Elèves des Lycées. Rapport Interne Laboratoire LIRDIS (Université Lyon-I) et Laboratoire CNRS-IRPEACS, 1990.
35. Larkin, J., Reif, F., Carbonell, J. & Gugliotta, A. :. FERMI: A Flexible Expert Reasoner with Multi-Domain Inferencing. Cognitive Science 12 (1), 101-138, 1988.
36. Moeschler, J. : Modélisation du Dialogue. Paris : Hermès, 1989.
37. Perrault, R. : An Application of Default Logic to Speech Act Theory. Intentions in Communication, eds. Cohen, P.R., Morgan, J. & Pollack, M.E. Cambridge Mass. : MIT Press, 1990.
38. Pierrehumbert, J. and Hirschberg, J. : The Meaning of Intonational Contours in the Interpretation of Discourse. In Intentions in Communication, eds. Cohen, P.R., Morgan, J. & Pollack, M.E. Cambridge Mass. : MIT Press, 1990.
39. Robertson, W.C. : Detection of Cognitive Structure with Protocol Data: Predicting Performance on Physics Transfer Problems. Cognitive Science 14 (2), 253-280, 1990.
40. Robertson, S.P., Zachary, W. & Black, J.B. : Cognition, Computing and Cooperation. Ablex Publishing Corporation : Norwood, New Jersey USA, 1990.
41. Roschelle, J. & Behrend, S.D. : The Construction of Shared Knowledge in Collaborative Problem Solving. To appear in C. O'Malley (Ed.) Computer Supported Collaborative Learning. NATO ASI Series, Springer-Verlag : Berlin, 1991, in press.
42. Schank, R.C. : Explanation Patterns. Lawrence Erlbaum Associates, Hillsdale NJ, 1986.
43. Schegloff, E.A. & Sacks, H. : Opening up Closings. Semiotica. 8 (4), 289-327, 1973.
44. Searle, J. : Speech Acts: An Essay in the Philosopy of Language. Cambridge : Cambridge University Press, 1969.
45. Sperber, D. & Wilson, D. : Relevance : Communication and Cognition. Basil Blackwell, Oxford, 1986.
46. Sycara, K. : Multiagent Compromise via Negotiation. In Distributed Artificial Intelligence : Volume II, Pitman Publishing, London, pp. 119-137, 1989.
47. Tiberghien, A. : Learning and Teaching at Middle School Level of Concepts and Phenomena in Physics: The Case of Temperature. In Learning and Instruction Vol 2,1, (eds.) Mandl, H., De Corte, E., Bennett, N. & Friedrich, H.F., pp. 631-648. Pergamon Press, 1990.
48. Toulmin, S., Rieke, R. and Janik, A. : An Introduction to Reasoning. New York : Macmillan, 1979.
49. Vanderveken, D. : On the Unification of Speech Act Theory and Formal Semantics. In Intentions in Communication, eds. Cohen, P.R., Morgan, J. & Pollack, M.E. Cambridge Mass. : MIT Press, 1990.
50. VanLehn, K. : Theory Reform Caused by an Argumentation Tool. Xerox Palo Alto Research Center Report P85-00102, 1985.
51. Vygotsky, L. : Thought and Language. Revised & edited by Kozulin, A. London : MIT Press, 1986.

33 [illegible] Mémoire de DEA de A. BESOMBES? Groupe Didactique de Physique, Université [illegible], 1990.

34 [illegible] Tiberghien, A.: Résolution de problèmes en physique par des élèves de lycées. Rapport interne Laboratoire LIRDIS (Université Lyon I) et Laboratoire CNRS-IRPEACS, 1990.

35 Larkin, J., Reif, F., Carbonell, J. & Gugliotta, A.: FERMI: A Flexible Expert Reasoner with Multi-Domain Inferencing. Cognitive Science 12 (1), 101-[illegible], 1988.

36 [illegible], 1988.

37 Perrault, R.: An Application of Default Logic to Speech Act Theory. Intentions in Communication, eds Cohen, P.R., Morgan, J. & Pollack, M.E. Cambridge MA: MIT Press, 1990.

38 [illegible] Discourse. Intentions in Communication, eds. Cohen, [illegible], Morgan, [illegible], Pollack, M.E. Cambridge, MA: MIT Press, 1990.

Analysis of Interfaces from the Points of View of Epistemology and Didactics

Andrée Tiberghien

Centre National de la Recherche Scientifique, Laboratoire IRPEACS, 93 chemin des Mouilles, BP167, F-69131 Lyon-Ecully Cedex, France

Telephone: (33) 72.29.30.18. Fax: (33) 78.33.33.70. Email: TIBERG@FRMOP11.bitnet

Abstract: We propose an analysis of the interface of learning environments in the domain of physics from an epistemological point of view. In particular we consider that the construction of meaning needs to take into account three main levels: theory, model and experimental field. Even in everyday contexts, we use a "theoretical positioning" to interpret the perceived facts. In that perspective, the distance between the learners' constructed meaning from the "forms of expression" at the interface (Hutchins & al.) and the meaning from the learning environment point of view which underlies its response to the learners' action is analysed.

Keywords: learning environment, epistemology, physics, mental models

1. Introduction

This paper is concerned with computer-based learning environments directed towards experimental learning sciences.

The present contexts of physics teaching and of the elaboration of learning environment are different. However in both cases there are epistemological positions concerning the knowledge to be taught which are not necessarily opposed.

Current teaching in France, at the levels of secondary school and at the beginning of university favours the set of formal rules of physics : a set of functional relations between physical quantities is constructed from a prototypical experimental situation. At the levels of elementary school and in middle school, another aspect is favoured : manipulation. In this case, the students construct an experiment, and, with the help of the teacher, they propose an interpretation which is often of a qualitative nature.

NATO ASI Series, Vol. F 86
Intelligent Learning Environments
and Knowledge Acquisition in Physics
Edited by A. Tiberghien and H. Mandl

In these presentations of physics there are underlying epistemological choices, even if they are implicit : 'since physics is an experimental science, it should be learned by doing manipulation" and "since physics consists of a huge part of formal knowledge in a mathematical language, teaching and learning physics models as such are essential'.

In these choices we find arguments linked to the nature of the knowledge involved and to the requirements of learning, and in particular arguments of the type: "in learning physics manipulation is essential because physics is an experimental science, and because learning is favoured by acting". In such arguments, the complex and multiple roles of experiment in science are not taken into account. For example, roles such as the opportunity of asking new questions, taking information, taking measurements, verifying, and the crucial case of validating are very seldom taken into account as a set of which the different aspects interact.

In the case of intelligent learning environments we may also find these basic choices and particularly the ideas of learning by doing. However the situation is different from usual teaching, in particular concerning the conditions of production of these environments and their use. Very few are introduced in usual teaching in particular at secondary level. The context of elaboration of learning environments is that of research, but the research disciplines concerned are often very different. One finds researchers in artificial intelligence, cognitive psychology, physics didactics (all of which may be considered as cognitive sciences). Consequently, very different learning environments are constructed. For example some emphasise the teaching of logico-mathematical operations in order to study the general processes of learning. In this case we often find the situation where the teaching content is only a support for studying logico-mathematical learning mechanisms (addition, subtraction, proportionality, ...). Other environments are more focussed on the teaching of specific knowledge in physics, where the structure of the physics knowledge to be taught plays a major role. According to these environments, the design of the interface, representations and possible actions will be different. In this paper we intend to give an analysis of such environments from the point of view of knowledge processing at the interface.

2. Importance of the interface and notions of distance

In terms of design, we share the point of view according to which the internal knowledge representation and the interface are strongly interdependent. (O'Malley, 1990). As a matter of fact, the interface conditions the learning situation - it is most often the place where students derive the information essential for their activity The interface is also the place where students may eventually interact and where they may give responses. Its nature therefore demands elaboration.

In other terms, we focus on the languages used at the interface to give information to the learner and on the languages which are to be used by the learner to process the situation. In terms of our perspective, this distinction is very important because we deal with the learning of a science of nature (physics). This discipline aims to interpret and predict the physical world with which we interact owing to our perception. This learning puts into play objects, events (movement, lighting, heating, ...) which can be very familiar since birth and which then correspond to representations which are not necessary those used at the interface. Moreover, work on students' conceptions shows how the gap between physicists' and students' interpretations of physical events can be different. Therefore the languages used (including iconic representation) at the interface are very crucial in the learners' construction of the meaning of the studied situation of the physical world.

With reference to Hutchins et al (1986), we make a distinction between the **ways of giving and obtaining informations at the interface** (information including knowledge on physics to be taught and on learner-environment interaction) and the **ways of action** on the environment by the learner. This distinction concerns the form of the output expressions and the form of the input ones at the interface. In their perspective this distinction is linked with the notion of distance "to the extent there are gulfs between a person's goals and knowledge and the level of description provided by the system". Norman (1986) proposes two distances :

- one between the interface display and the user's construction (goals, representation of the situation, plans, ...) from the output of the system through the interface which is called the "gulf of evaluation", and
- the other between the user's intentions and expectations, and the interface mechanism leading to the system answer to user's actions which is called the "gulf of execution".

These notions of distances are also relevant for physics learning situations which put experiments into play. Most of the work on conceptions shows that the interpretations given of a physical system (experiments) by a student or by the teacher (as representative of the physics teaching) can be completely different, in that sense there is a distance which corresponds to the evaluation gulf and that we call in our case distance of meaning. We keep the same word "execution" for the other distance. In this paper we present our theoretical approach to these distances, mainly "distance on meaning". Then we discuss it in the case of learning environments and formulate a grid of analysis for the interface of learning environments.

3. Modelling as a reference in knowledge processing

To analyse these distances we take a same point of reference which should be relevant both to the learners' point of view and to learning environment interfaces, and also be in keeping with the aim of the learning environments : learning a science of nature and more particularly physics. Our choice bears on a process of knowledge and not on knowledge itself : this is modelling.

In modelling we consider three levels, theory, model, experimental field. The theory gives meaning to the modelling, and in this sense it is relevant for both the learner and for physics. We consider that theory corresponds to the explanative system and that it includes causality. This supposes that all mental activities, even if they are inefficient, are not aimless - an explanation can be found. Moreover, with epistemologists of science, we consider that a concept does not come from the extraction of factors common to a number of observations, but rather requires an intellectual positioning (Cassirer, 1977) in which the choice of a theory intervenes. Even for non-scientific knowledge, for example for everyday knowledge processing "a theoretical positioning" intervenes. For example, it is well know that we see what we want, we process only a part of information coming to our senses. This processing supposes what we call a theory. This theoretical positioning gives meaning to perception, it is strongly linked with the type of explanation of the situation given by the actor and then with the type of causal links established. In this perspective, we consider that it is crucial to analyse the type of causality used by learners to the extent that it implies the meaning given to a situation.

In the following we focus mainly on the level of theory for the learners' construction and in physics to the extent that it conditions the construction of meaning.

3.1 Mental models and learners' causal approach

In our previous work in physics education concerning students' conceptions, we have shown that one can interpret an important part of their interpretations in terms of unsuitable use of causal relations. This leads to the claim that learners' theories plays a crucial role in their mental models.

In order to analyse this causality, we propose Aristotelian causalities as a reference : material, efficient, formal and final (Kuhn in Bunge et al 1971).

- material : for example this type of causality is used when students consider that cotton, wool are hot, or heat "because it is wool", i.e. it is a property of the wool.

- efficient : for example this cause is very often involved when there is a change. A battery lights the bulb ; the cause of the lighting up is the battery or a property of the battery (energy, electricity). Or, a ball hits another motionless ball which starts moving : the movement of the first ball (or the kick given to it) is the cause of that of the second ball....

- formal and final : these latter two are often used together and we do not try to distinguish them systematically. For example, when students invoke the function of an object to predict or interpret (because the coffee pot is made of metal, metal will keep something hot), we consider that it is a mixture of these two causes.

In the case of electricity, the research work concerning students between 9 and 14 years old (Tiberghien et al 1976 ; Ben Hamida 1980 ; Osborne et al 1985) shows that even young students propose an interpretation of the functioning of simple electrical circuits.
The method of investigation used is that of individual interviews. These start off the student and provide the necessary materials - battery, bulb, wires - and ask for the bulb to be lit up. If the student does not succeed in lighting up the bulb, after a certain time, the interviewer helps. When the circuit functions, the interviewer poses a question of the type : "how does that work?".

Here are three typical response to this question :
"The electricity passes by the contacts of the battery and it arrives in the wire, that's what gives the light .... the electricity comes from the battery by the contacts"

" The battery is used up after a while because it gives out, it gives out but if we disconnect it there's going to be electricity in the battery"

"The bulb lights up .... because there is electricity"

These examples show the use of linear causal reasoning, of the "efficient" kind, in the sense of Aristotle, with the battery as cause in the first two cases and the electricity in the third.

This efficient causality is used very often when there is an abrupt change just as much in electrical circuits (lighting up or extinction of a bulb) as in the case of movement (a blow leading to an immobile object being put in motion). There is nevertheless a large difference between the case of movement and electrical circuits in terms of the use of a mediator : the case of heating or of cooling is close to that of electrical circuits in terms of the mediator and can allow us to go deeper in our interpretation.

We interpret these differences by the invocation of a mediator by two characteristics of the situations interpreted :

- the **spatial arrangement** of the objects linked to the cause and effect.
In the case of electrical circuits, the cause or generator is distant from the effect (lighting of the bulb, heating in resistance) with a link materialised by electrical wires.

In the case of heating situations with an intermediary material between the source and the place where the effect is observed, the students use causal reasoning with the source as cause, the object heated as effect, and a mediator between the two which has also the property of displacement from the source of the object heated. On the other hand, when source and object heated are in direct contact, the invoking of a mediator is much more rare.
In the case of movement, the objects (cause and effect) are in contact at the moment of the change. For example, a ball in movement which hits a fixed ball, or a ball which is released.

Thus in these first two cases, there appears to be an **intellectual need** for a mediator which "transmits the cause". In effect, the mediator, often named as "heat" by the students, or "hot air" for heating situations and "electricity" or "energy" for electrical circuits, does not correspond to a direct perception, and furthermore is evoked with its own particular property - that of displacement. In this sense it is different from the source, the effect and its support.

- **evolution in time** of the role of the objects linked to the cause.
In the case of movement, the object which has been the cause (support, other ball) becomes useless : the cause of the continuation of the movement can therefore only be situated in the object in movement itself. In the case of heating and electrical circuits, the cause (battery, burner) remains the cause of the heating or lighting up of the bulb. It is constantly at the origin of the mediator. This can explain why certain students consider electricity or heat as the cause. One can note that this difference shows the importance of analysing to what extent the students perform a reading at a given moment on a duration, or do not differentiate these two readings (Tiberghien et al, in press).

3.2 A cognitive model of young students with respect to batteries and bulbs

In what follows we describe a hypothetical model corresponding to a cognitive structure constructed by young students in their approach to simple electrical circuits. For us, the term "cognitive model" corresponds to a construction of the *researcher*. It permits the interpretation, and in certain cases the prediction of the productions of students.
In order to elaborate the cognitive model we begin from elements of the causal structure. In these interpretations the students, the battery, the bulb and the "current" (mediator) have the following properties :

The battery (cause)
- gives "electricity" ; this 'giving out' is a characteristic of the battery
- is used up after a certain time.

The bulb (effect)
- uses the "electricity", or "consumes" it
- gives light
- can heat (in certain cases).

The "current" (mediator)
- displaces itself from the battery to the bulb.

The interpretation and the prediction are based on a causal relation linked to the action of lighting up (figure 1).

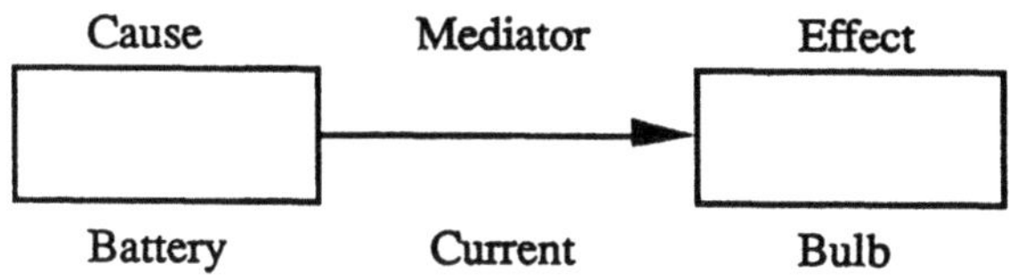

Fig. 1. Energy chain for battery-bulb situation.

3.3 Role of causality in the interpretations of students

Several researchers (Closset, 1983, Shipstone 1985) have proposed the notion of "sequential reasoning" in order to interpret the responses of high-school and university students concerning electrical circuits. This consists of using a physical quantity called "current" which will have the properties of a fluid in the sense that it will not be modified only by the elements traversed. Thus what is upstream does not affect what is downstream.

Even students after several years of university studies in physics use this kind of reasoning. Our analysis in terms of causality allows us to propose that this reasoning is compatible with the causal structure presented below, with respect to young students (table 1). The generator (cause) provides a constant current (mediator), and the bulb or the resistor "consume" this "current" (mediator). A single variable is used - the "current" (mediator) - which depends on the arrangement in space of the elements - what is upstream does not influence what is downstream.

This causal reasoning is close to a kind of everyday reasoning used for interpreting actions. They have a common characteristic - the use of a very general linear causal structure to deal directly with particular cases, in order to predict and/or justify actions. This reasoning is

thus very local. We make the hypothesis that this type of causal reasoning forms part of common knowledge. It is at the same time very effective in practical everyday matters, and of a very low 'cognitive cost' : a cause, an effect, a mediator. Thus this reasoning observed in young students in the case of electrical circuits is **frequently used by adults in the practical field of everyday life** (Schiele 1984).
We therefore make the following conjecture : in the case of students at the Masters level, the use of sequential reasoning corresponds to the mobilisation of a causal relation which is not relevant, but which is available because it is constantly used in everyday life with respect to similar concrete situations. We interpret this mobilisation of causal reasoning in terms of :

The gap between the good mastery of the rules of models of physics by students (such as those of electrokinetics) and their miscomprehension of the meaning of these models, in other words, their miscomprehension of the theoretical level.

The distances of meaning and execution allow these miscomprehensions to be interpreted. As a matter of fact, in their interpretation the students used the entities given in the question which are very often physical quantities and they give to these the same properties as they use to interpret real situations in everyday life, and in particular they use causal reasonings. They do not take into account the "long detour" or the distance between the languages of processing and that of description in terms of events or actions. From another point of view, we find like Greeno (1989) the importance of the difference between concrete and abstract entities and their specific role in semantics.

This analysis can be transposed in the case of learning environments. The entities used at the interface and the causal relations which are behind their behaviour (implicit or explicit) play a crucial role in the construction of mental models and then in learning. These entities and the explanative system behind them have to be compared with those of the learner.
We recall Larkin's results where the entities given in the problem statement are taken as such by the novices to solve it instead of restranslating them in another level of description (Larkin, 1983). A deep difference between novices and experts in problem solving is in the phase of "representation of the problem". The novices have a short phase of reprocessing of the presentation of the problem and process it **with the same entities** as are used in the text, whereas the experts take time for reprocessing the problem and introduce **several levels of description which implies different entities** usually more conceptual .

4 The case of physics

In physics, the nature of causality is an object of debate. We give below some elements to show how it is important in the functioning of physics knowledge.

Laplace argued for a deterministic causality : "we must thus envisage the present state of the universe as the effect of its previous state, and as the cause of that which will follow" (1825). **This causality is very elaborated** : it does not signify that in a functional relation one establishes a causal relation between the physical quantities. Thus, for example, Ohm's Law (U=RI) does not lead to saying that voltage (U) will be the cause of current (I) (or reverse) or in the fundamental mechanical relation that force will be the cause of acceleration (or the reverse). This is a case of what Halbwachs calls "circular causality" (1971). This does not of course exclude the case where simple causal reasoning can be used when the independent variable is fixed.

In order to consider causality, physics introduces

> "the notion of the physical system, that is to say a certain set of objects ... that we know how to isolate by thought and to distinguish unequivocally from the rest of the physical world ... noting that this isolation is only done by "thought" and that the system is in fact subject to influences and actions on the part of the rest of the Universe. It is precisely this which we aim to study. ... One must in general **place oneself in the framework of a certain model and constitute a physical theory ...**
> (Halbwachs, 1971)

Thus, in the framework of mechanics, one characterises a system by its movements and its parts, and by the forces which they exert on each other, even though the same system, studied from the point of view of thermodynamics, will be besides characterised by its temperature. Determinism consists in proposing that knowledge of physical quantities giving the state of the system at a given moment allows the anterior and posterior states to be known. Prediction is possible.

Since Laplace, the notion of causality has evolved and is currently the subject of debate (Le Débat, 1990). Other positions appear, such as that of Prigogine and Stengers (1979): "We find ourselves in an irreducibly random world, in one where reversibility and determinism are specific cases in point, where irreversibility and microscopic indetermination are the rule".

It is important to note that the causality of Laplace is not eliminated but becomes a specific case. It is this case which underlies most of current physics teaching, from elementary school to the first years of university.

This debate shows the importance of theoretical choices on causality in conceptual construction; they condition the meaning of the relations between entities which allow the interpretations of the situations studied. This causality is very different from that used most often in everyday life - the latter being close to the efficient causality of Aristotle.

This short presentation shows that in the case of physics, the interpretation needs a 'long detour' from the real objects. This detour is a modelling of the real situation in terms of different **conceptual entities** which implies specific languages with given syntaxes. For example, the interpretation of an electric circuits usually needs a representation in a two dimensions space: the electrical diagram, a algebraic representation en terms of physical parameters, and real objects (lamp or resistors, battery, ...) or their representation in terms of events (lamp heats, is lit up, ...) this last representation has no conventional code but is necessary, at least in thought, since the theoretical construction aims to predict or interpret events. Therefore, **interpretation in physics needs several levels of description and therefore several types of entities**, at least real objects and physical quantities and very often different level of physics conceptual entities (categories, system, physical entities, principles, ..).

This point, very important for the point of view of physics, needs to be taken into account when designing interfaces. It will be on this basis that our analysis grid will be constructed.

In this modelling approach we retain the following points :

1. theoretical construction (model + theory) is not extracted from experience (opposition to positivism). The history of science shows well that observation and manipulation alone are necessary but insufficient to establish a law or more generally to establish a theoretical construction, even if it plays an important role.
2. Theory gives a meaning to the functional relations between physical quantities. Thus, causality forms a part of the theory.
3. The theoretical constructions are not absolute, at least from two points of view :

° they evolve in time

° they are only validated on a restricted experimental field. There is no universal theory : knowledge is local.

4. The theoretical construction is validated owing to experimental situations (coherence between different parts of theory are also part of validation processus)

It is interesting to note that the modelling approach in experimental sciences which emerges explicitly between Galileo and Newton is much more recent than the demonstration in mathematics (Cohen, 1987).

An interpretation of this long distance in time between these two scientific processes (modelling and demonstration) could be that it is much more difficult from a cognitive point of view to deal **both** with real objects and conceptual entities with necessity of validation process in between.

We consider that these different levels of description **cannot** be avoided and in physics the "feeling of directness" cannot be applied without a deep analysis.

The notion of distance introduced by Hutchins et al. in the case of physics, allows to take into account the huge distance between the input level of description which can be the material situation and in that case is very near to the perception and the levels used for treatment at conceptual levels with entities having characteristics which are different from the real objects :

- there is no necessary biunivocity between entities and objects of the situation (an object can be described by several physical quantities or an entity can correspond to the whole system) ;
- the causal relations between entities do not obey the same laws;
- the syntaxes can be different (for example geometrical or algebraic).

5. The case of learning environments

5.1 Situation and *enjeu* [1]

A learning environment can be used in diverse contexts, and in particular it may be an integral part of teaching or be used as a supplementary aid, left to the free disposition of the students. In the vast majority of cases, and the cases which we assume, the learning environment will be used by one or two students within the framework of teaching even if this is not during classes. That is to say that the purpose of the computer-based environment is to facilitate learning of some part of the knowledge taught in physics within the school or university framework. In the following we suppose that the learning environment will be used in the case where *the enjeu is the learning of the taught knowledge.*

5.2 Importance of form of expression

As we proposed, we take the situations where the aims of the learner in interaction with a learning environment is to learn physics in order to acquire a deeper physical meaning of the material world. This proposition requires us to make precise what we mean by "construction of meaning" in learning physics. The meaning is linked with the way in which the thought is made explicit. We refer here to Lavoisier (1743-1794) : "The impossibility to isolate the nomenclature of science and science of nomenclature depends on the claim that any physical science is necessary composed of three things :
- series of facts which constitute science;
- ideas which recall them;

---

[1] "Enjeu" = what is 'in play', 'involved', or 'at stake'. We retain the French word here in view of its multiple meanings.

- words which express them.....
however certain the facts may be, however right the ideas which (the facts) would have be given birth, may be, they would transmit only wrong impression if we did not have right expression to give back (render) them.".
In this perspective, "construction of meaning" implies to take into account both the discipline (content of knowledge) and the learner.
In our approach we take as a basis a very general proposition that language (words, symbolic representation, syntax) used is crucial in the "learner's construction of meaning". Morevover, the language allow to analyse knowledge processing involved in the teaching-learning situation created by the learning environment. Several types of knowledge are involved with different processing and with possible different languages
- physics to be taught as it is written in computer language;
- physics to be taught as it is presented at the interface;
- learner's construction of mental models from information taken at the interface and possibly from real experiments;
- learner's answer formulated at the interface;
- computer learning answer.

The case of learning environments gives a very rich opportunity to manage the language used in the teaching-learning situation (to the extent it helps to distinguish between the language as tool of representation and the language as tool of communication (Laborde, 1982)). Constraints of learning environments lead to make explicit the forms of expression concerning :
- physics as it is taught at the interface (and consequently the output of the systems) and
- ways of acting or/and answering that the learner can give (the input to the system).

5.3 Role of the different forms of expression

Whereas Hutchins *et al.* used these notions in reference to a computer system, we extend this notion to the cases where the interaction of the person is also with learning environments in physics. This extension allows us to use it in analysis interface of learning environment in physics and even extend in the case of manipulation with real objects is involved.

The learner's goals will depend on the type of problem the learner has to solve. We consider only the cases where at least a part of the screen presents a simulation, a data treatment on line or a microworld (constructed from a model in physics). In that cases, there is a perception of events at the screen (or in reality). The events can happen with representation of real or conceptual objects (e. g. impact of balls, or deformation of a graph). In the case of real objects, perception will depend on the way of representations of objects (faithful, symbolic, ...) and then the level(s) of description used by the learner to make explicit what he perceives, will depend on the representation and on his goals. A difference between his level of description and the representations can happen, we have then a **distance** between his thoughts and the output of the learning environment.

If it is up to the learner, the level of formulation of his answer will be very different according to the level of formulation of the problem and the level of description and interpretation of the perceived situation. If the learning environment imposes a language we need to introduce the notion of **distance** between the learner's thought and the level of language imposed by the environment to give an answer or more generally to act on it at the interface.

Then, we have two types of distances which correspond to the gulf of evaluation and the gulf of execution proposed by Hutchins *et al.*

5.4 Analysis of distances

First of all, we need to make precise the distances in the case of physics learning environments. In the figure 2 we present the two distances.

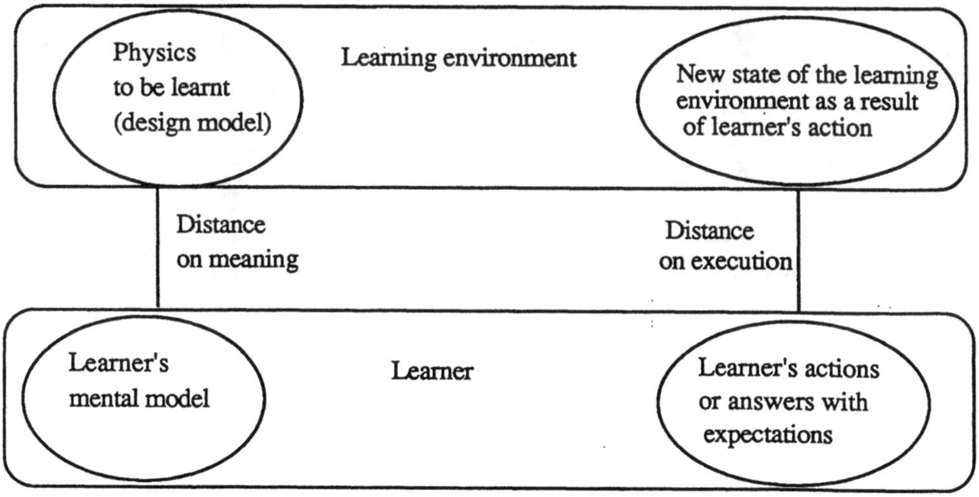

Fig. 2. Distances of meaning and of execution.

These distances suppose that we consider that the learner constructs a mental model from his/her interaction with the learning environment. The first distance corresponds to the gap (or the gulf) between this model and the model presented at the interface and the second one corresponds to the gap between the learner's expectation of his/her actions and the system answer. Note that the learner's model module in learning environment intends to reduce these gaps, however the question of distance is actually still relevant.

This approach with the notion of distance, takes into account the form of expression or in other terms the symbols. Our theoretical approach based on an epistemological analysis of physics and psychological analysis with mental models converges with Greeno's approach (1989), in both cases we intend to analyse knowledge processing "as a relation between an individual and a social or physical situation".

In figure 3 we present the diagram with our interpretation in the case of physics learning environment of :

- the input (from learner's point of view or output from learning environment point of view);
- learner's treatment of the situation;
- the output (input from learning environment point of view).

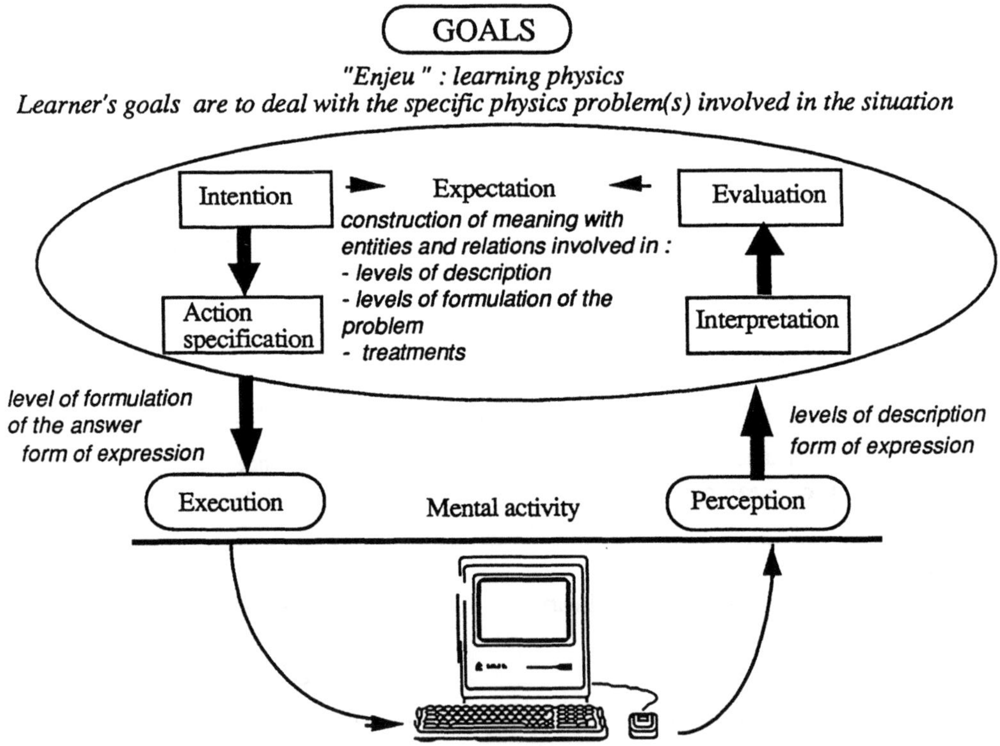

Figure 3. Input-output in a learning environment

In the perspective of learning physics, the question of minimising the distance has to be taken in different terms for the case studied by Norman and Hutchins because, this minimisation corresponds to having the lowest cognitive effort. In addition, the difficulty in interpreting physical situations is that cognitive effort is worthwhile if we want to use the same theoretical construction for a wide variety of situations ; if not, the everyday approach is more relevant and consumes less cognitive effort. As we stated, it is easier to have very general reasoning and apply it directly to available objects.

6. Grid of analysis of the type of representations chosen in designing the interface

According to our choices of modelling as reference, the different forms of expression used to present a representation at the interface (figure 2) can correspond to one of the three levels - theory, model and experimental field - or to the relations between the levels. Our grid intends to characterise this. We present first our items by levels : experimental field of reference, model, theory and then we discuss the relations between them.

6.1 Levels of the experimental field

In a physics modelling process, the experimental field plays several roles:
- it is the field of validation of the theoretical construction and then its **extension** has to be taken into account;
- it is the place where information is taken by direct perception of events or by measurements (which are specific events), the type of representation used which necessarily makes selections from the real material situation, is crucial. Then we consider the **gap** between the real situation of reference and the represented one at the interface bearing on :

    the difference in possible perception of objects, events and their properties;

    the form of expressions used for the representation.

In a learner's modelling process, to the extent that a situation of physics can involved familiar situations, the learner's previous knowledge of them plays a very important role. The level of experimental field is essential because the perception of the material environment plays

a crucial role in learner's theories or mental models construction (Vosniadou, Bliss & Ogborn, ibid). Consequently the form of representation used at the interface will depend on :
- the learners' mental models which he will construct of the situation and
- the relations that he will establish with his own theories of similar situations in the real world and then the construction of his learning.

6.2 Extension of the experimental field

The notion of experimental field of a theoretical construction is linked with the validation process. It is the set of experiments for which the theoretical construction is valid. Usually, for many reasons such as duration of teaching, students' initial knowledge, students' interest, ... in the teaching content only a few and sometimes one prototypical experiment is used both to introduce and to verify the theoretical construction. Note that this process is different from the scientific research case. For reasons similar to the case of usual teaching the learning environments cannot respect as such the research process in physics which itself is not studied by a researcher alone. Consequently we need to ask the following questions :

- Are the possible real situations simulated at the interface valid for the theoretical constructions involved in the learning environment; if not what are the means to invalidate the theoretical construction (for example predicted events do not occur) ?
- Is the limit of validity of the theoretical construction in play is treated as such ?

6.3 Gap between the real situation of reference and the represented one at the interface

As we have discussed before, the theory plays a crucial role in the selection process of the aspects taken into account in a situation to studied. In the case of learning environment, the criteria to select the aspects to be represented which are at the theoretical level should be explicit at least for the designer; the following questions should be asked :

- What are the reasons for selecting the objects ?
- What are the reasons for selecting such properties of objects ?

The drawings used for the representations play a crucial role in the selection. Then new questions :

- Is the representation only with drawings or photography or video or does the representation also use some specific entities coming from:
  - the usual computer interface ;
  - a design specific from the given learning environment
  - physics (electrical circuit diagram, vectors representing velocity, force, ...).?
- Does the representation use at a given time one type of entitites or several ?

If we take, for example the case of the learning environment SUOMO (Hron, Mandl, ibid), the situation is a selection of the objects and events: there are only two persons on the conveyor belt, the surroundings on the conveyor are not represented, ... and the representation adds several types of entities, on the "conveyer" there is reference points named X. Y. Z. ... on the conveyor frontiers , ...

In the case of Graph and Tracks (McDermott, ibid), the representation is a drastic selection of the material situation only the objects relevant to the theoretical constructions are represented.
In the situation of Reiman (ibid), no material objects are represented, some are only evoked by text : light source.

6.4 Familiarity of real situations of reference in the interface representation

Different cases may obtain :

- Objects and events are usually perceived in our environment
This case can happen even if the situation is constructed and is not extracted directly from the environment. For example an electrical circuit with batteries, wires and bulbs are not usually assembled as presented but the situation put into play usual objects and events (lit up, heating of the bulb or the battery, ...).

- Real objects and events are not usually perceived in our environment.
As a matter of fact the learning environment can give the opportunity to be aware of inaccessible experiments for several reasons :
- security (crash of cars, ..)
- economical cost
- inaccessibility (distance, ...)

The extension of the field of experiments can play an important role in learning. The following questions can be raised :
- Is the link with usual perceived events made ?
- Are the involved objects known ?

- Real objects but not events which are usually perceived in our environment.
For example a world without friction, in that case objects into play can be familiar but do not behave as usual; the events are different. Therefore the following question need to be asked :
- Does the learning environment give the possibility to simulate also events really perceived with the same material objects ?
This question raises the problem of the field of validity of the theoretical construction :
- If the learning environment gives the possiblity to simulate also events really perceived (in our world with friction for example) does it give the possibility to verificate that the theoretical construction valid for a given modelling (nonfrictionless world for example) is no more valid for our world ?

- Non-visible objects and non-usually perceived events.
Multiple cases are possible particularly when we change in scale in time (what is going on in less than a second or more that 100 years), in distance (less than a millimetre or more that 1000 km) or in time X distance space. If we take for example the case of a microscopic world (distance scale) a question is raised :
- Does the macroscopic situation which corresponds to the microscopic one can be represented? is it evoked ? if yes with what forms of expressions ?

This question raises the domain of validity of the theoretical construction. For example, a microscopic model of matter has specific properties of interaction between particles which are not in the usual world. In that case the relations with the macroscopic world are complex (Meheut, 1990).

Possible actions of the learner
From the previous questions, it is easy to analyse this aspect.
- On what type of entities can the learner act ?
  entities representing faithfully the objects (moving a ball, ..)
  entities specific from physics (and then belonging to the theoretical construction)
  entities specific from the learning environment.

- What are the responses of the learning environments
  the representation of the real situation is modified
  the represention using entities of physics are modified

the representation using specific entities of the learning environment are modified.

the representation of the real situation and that at the level of theoretical construction are simultaneously modified ?

In the case of Suomo (Hron, Mandl, ibid) there is no direct access to the entities, but the drawings are elements of validation, the events being simulated.
In the case of Graphs and Tracks (MacDermott, ibid) only the ball and the track are represented, that is only the pertinent objects for the physics knowledge into play are selected. The learner can modify in one case the representation of the real situation by adding or subtrating plots linked with the orientation of the track according to a representation at the level of the theoretical construction (graph of the functional relationship between physical quantities (distance, velocity, acceleration, time)) and in another case the reverse.

7. Levels of model and theory and relations with the level of the experimental field

Usually the theoretical level is not taught as such, that why we treat the theoretical construction (theory and model) as a whole.

As we have shown before, in the physicists' and the learner's approach, the constructed models depends **both** on the theory and on the situation.

In this grid we insist on the following aspects :
- the coherence between the diverse aspects of the model
- the coherence between the model and the experimental field which implies taking into account

the status of the entities used at the interface at the model level and their relations with the experimental field level.

the possible roles of interpretation and prediction of events played by the theoretical construction in the learning environment.

In this analysis, the notion of coherence is central. The meaning of the entities will be very different according to coherence is considered internal to one level (model or experimental field) or also in the relations between the different levels

7.1 Coherence at the level of the model

In this paragraph we consider the case of internal coherence at the level of the model. The microworlds are good cases for analysing this coherence. As a matter of fact, a microworld, with its primitives and rules, is a coherent world in itself. It could be considered as autonomous knowledge to the extent that its coherence does not need an external confrontation. As an extreme we could say that the world presented at the interface could be "autosufficient".

- Are the entities or forms of expression used specific of physics models or specific of the microworld ?
- Do the entities belong to these several types (a model of physics, specific of the learning environment, ...) in a same representation or in different ones (different screen pages or different windows) ?

An example is that of "Refract" (Reiman, ibid) which consists of the entities and rules of the model of geometrical optics. In that case, the events are not those really perceived but are "conceptualised events" which are transformed in real events perceived at the interface. This world used two types of entities or forms of expressions : geometric and algebraic which belong both to the geometrical optics model. With Hutchins and al (1986), we consider that the world of interest is explicitly represented and in this sense there is no intermediary between user and world.

In microworlds, the learner can act directly on the entity of the world. The possible actions have consequences (create events) at the **same** level, that is if the entities are at the level of a physics model, **the learning environment answer will be still at this same level and not at the level of experimental facts.** This is the case of Refract (Reiman, ibid) : the entities belong to the model of the optical geometry, the learner can act on them and have an answer in terms of the same type of entities.

7.2 Coherence between model and experimental field

In the case of the model is presented in relation with its experimental field, we can take again the same analysis as previous for the microworld but we need to consider also to what extent the coherence between model and experimental field, what we call "semantic coherence" does affect the internal coherence of the model. When coherence is established between these levels, then a **theory is necessary to give meaning** to the relation between these levels (a same model can be used in different theories, but it gives different meaning to the experimental facts).

In this case several types of forms of expression are necessarily used. Some representing the objects and events of the experimental field and some specific of physics (at least of physics teaching). As in physics there is not necessarily correspondance term to term between these two types of entities. The coherence has to be at the level of interpretation and prediction of facts. This type of necessary coherence is a big constraints but also can give the possibility to adapt the physics taught model to the learner's mental model with two conditions :

- the coherence between the model in question and the chosen experimental field had to be respected. This leads usually to having a restricted experimental field;
- the model respects the basic physics principles (for example conservation ones, ...).

In these articulations between the different levels, many possible meanings can be constructed by the learner, therefore the theory level is involved in the reduction of the distance. Two main questions are raised :

- What is the learner's meaning of the forms of expression that are used to present the model?

  Does this meaning needs to include other levels of form of expressions both at the theoretical level and at the level of objects and events ?

  Can this meaning be constructed mainly from information given by the environment ?
- Is there necessity to construct meaning to relate the knowledge at the level of model to knowledge at another levels ?

These types of questions lead to taking into account the compatibility between model of physics to be taught and learners' mental models.

In that perspective, we can have simulation which is run by the conceptual models close to learners' mental models. This type of learning environment may offer a possibility of a certain type of validation of the mental models. That supposes that the entities used by the learner to interpret the material situation and those used in the simulation are compatible with a theory and particularly with the causality used by the learner.

An environment like Quest (White & Fredericksen, 1987) comes into this category. Concerning electricity the authors consider that "electricity cannot be understood within a single model or perspective; rather, electrical systems must be modelled at multiple levels for knowledge to be robust, and to be effectively applied to new situations". They proposed models of different orders ruled by different causal relations (in our terms there is a change at the theoretical level) and which intend to match with learners' mental models.

In the case of Graphs and Tracks (McDermott, ibid) different graphs are represented (one type of entities), the learner can modify them, the answer is in term of **another** type of entities : those of the material situations. The relations between levels of model and experimental field is the main cognitive activity demanded by this learning environment. A possible validation can be done and then the construction of meaning of the objects and events (experimental field level) through physics is involved.

These questions are intended to assist in the analysis of the learning environment in the perspective of the type of knowledge which is possibly processed during the interaction learner-system. It also can lead to analysing the adequation between the aim of the learning environment and the type of knowledge which is into play.

References

1. Ben Hamida, J. : Modèles de fonctionnement de circuits électiques chez les enfants de 12 ans. Thèse de 3ème cycle, Paris, Université Paris 7, 1980.
2. Cassirer, E. : *Substance et fonction. Eléments pour une théorie du concept.* Eds de Minuit : Paris (1st edition in German 1910), 1977.
3. Closset, J.L. : Le raisonnement séquentiel en électrocinétique. Thèse 3ème cycle, Université Paris 7 1977.
4. Cohen, B. : Newton et la découverte de la gravitation universelle. Pour la science. pp. 101-110. 1987.
5. Duit, R., Jung, W. & Rhöneck, C. v. (Eds) : Aspects of understanding electricity. Proceedings of an international workshop, Ludwisburg (IPN Kiel). 1985.
6. Research on physics education: Proceedings of the first international workshop (La Londe les Maures) Paris : Editions du CNRS 1984.
7. Greeno, J.G. : Situations, mental models and generative knowledge. In D. Klahr & K. Votovsky (Eds). Complex information processing. Hillsdale, N.Y. : Lawrence Erlbaum. p. 285-318. 1989.
8. Gott, R. : APU Science, Electricity at Age 15. (Department of education and science, London). 1984.
9. Halbwachs, F. : Causalité linéaire et causalité circulaire en physique. In Bunge,M.

Halbwachs, F. Piaget J. & Rosenfeld, L. Les théories de la causalité. P.U.F. Paris. p. 42-43 1971.

10. Hutchins, E.L. Hollan,J.D. and Norman, D.A. : Direct manipulation interfaces. In D.A. Norman and S.W. Draper (eds) User centered System Design, Lawrence Erlbaum, Hillsdale, NJ. 1986.
11. Johsua, S. & Dupin, J.J. : L'électrocinétique du Collège à l'Université. Bulletin de l'Union des Physiciens, n° 683 p 779-799. 1986.
12. Laplace, : Essai Philosophique sur les Probabilités, In Le Strat (Ed) Epistémologie des sciences physiques. Nathan. 1990
13. Laborde, 1982 : Langue naturelle et écriture symbolique. Deux codes en interaction dans l'enseignement des mathématiques. Thèse d'état. Université de Grenoble. 1982
14. Larkin, J. : The role of problem representation in physics. In D. Gentner & A.L. Stevens (Eds). Mental models. Lawrence Erlbaum, Hillsdale, NJ. p.75-98. 1983
15. Le Débat : La Querelle du déterminisme Gallimard.1990.

16. Méheut M., & Chomat, A. The limits of children atomism; an attempt to make children build up a particle model of matter. In P.L. Lijnse, P. Licht, W. de Vos & A.J. Waarlo (Eds). Relating Macroscopic phenomena to microscopic particles p. 266-282. Utrech CDB Press. 1990.

17. Norman, D.A. In D.A. Norman and S.W. Draper (eds) User centered System Design, Lawrence Erlbaum, Hillsdale, NJ. 1986.

18. O'Malley, C. : Interface issues for guided discovery learning environements. In Elsom-Cook M (Ed) Guided discovery tutoring. p. 24-41. Paul Chapman Publishing Ltd : London. 1990.

19. Osborne, R. & Gilbert, J. (Eds).: Some issues of theory in science education (University of Waikato, Hamilton). 1985.

20. Prigogine, I. & Stengers, I. : La Nouvelle Alliance. Paris, Gallimard, p. 23. 1979.

21. Schiele, B. : Note pour une analyse de la notion de coupure épistémologique. In Communication Information, vol VI, n° 2-3 p.43-98. 1984

22. Shipstone, D.M. : On children's use of conceptual models in reasoning about current electricity . In R. Duit, W. Jung & C. v. Rhöneck (eds), Aspects of understanding electricity. Proceeding of an international workshop, Ludwisburg 1984 (IPN, Kiel), pp. 73-93. 1985.

23. Tiberghien, A. & Delacôte, G. , Manipulations et représentations de circuits électriques simples par des enfants de 7 à 12 ans. Revue Française de Pédagogie, n°34. 1976.

24. Tiberghien, A., Psillos, D. & Koumaras, P. : Physics instruction from epistemological and didactical bases. to be published.

25. White, B. Frederiksen, J. : Causal model progression as a foundation for intelligent learning environments. BBN Report No 6686. Cambridge,MA. 1987.

4. Design of Learning Environments

Computer Simulation of Historical Experiments and Understanding of Physics Concepts

L. Borghi

Dipartimento di Fisica "A. Volta", Università di Pavia, Italy

Abstract: This paper deals with an approach to the teaching of physics which uses computer simulations to reconstruct significant experiments from the history of physics. Simulations are designed to assist students in working out ways of reasoning about the problems which engendered the original experiments. Focusing on research findings about the understanding of concepts of velocity and acceleration, we based computer simulations on Galilei's experiments with an inclined plane. We prepared worksheets tailored to induce students to reflect on fundamentals of kinematics. In order to improve the learning environment we recently accompanied computer simulations with sound, images and animations.

Keywords: computer simulation, historical experiments, Galilei, kinematics, multimedia approach

1. Introduction

Findings of research [5] show that a crucial point in learning physics is that students focus their attention on their own thinking and compare their explanations with accepted physical theory. In the belief that this comparative activity could be enhanced by analyzing, at least for some relevant physics subjects, the process of theory construction we think it useful that students analyze historical experiments.

Modern historiography of physics offers a wealth of historical case studies, but usually it considers only the theoretical side of science, i.e. analysis of original sources such as papers or treatises. It would be important to provide a historical laboratory to complement this analysis but original apparatuses are not available and accurate reproductions are difficult to set up. Computer simulations can help to overcome some difficulties and lead students to understand how original apparatuses work and to gain insights of the environment in which the experiments were carried out.

NATO ASI Series, Vol. F 86
Intelligent Learning Environments
and Knowledge Acquisition in Physics

We developed a teaching strategy based on computer simulations which present historical experiments and guide students in their analysis. Our aim is to suggest the pattern of reasoning which inspired the design of experiments.

Such historical analysis, in which the students are given the opportunity to personally reconsider original experiments, can play an important role in supporting the understanding of fundamental physical concepts especially when the historical experiments were designed to find the significant variables in a phenomenon and their relationships.

Focussing on findings about the understanding of concepts of position, velocity and acceleration [3-4], we chose to reconstruct the Galilei's experiments with an inclined plane and prepared a package dealing with the fall of bodies for use in introductory Physics courses. The simulations assist students in working out ways of reasoning about the problems which engendered the original experiments.

The debate on Galilei's work is wide and far from reaching unanimous conclusions, but we decided to follow the Drake's reconstruction [2]. Galilei carried out experiments with little balls moving along an inclined plane in order to find the relationship between displacement and time. (As it is well known, he was convinced that the analysis of the motion of objects on the Earth could lead to describe the motion in the heaven.)

The simulations present the sequence of his experiments: they are designed to reconstruct the original experimental situation and to help students compare their own way of reasoning with the model underlying each simulation.

The package includes computer programs (implemented on IBM compatible personal computers), a teacher's guide, and a student guide with worksheets. In addition to suggestions for the use of the programs, the guides contain a historical section with relevant original sources. The teaching material has been described in detail elsewhere [1].

In the following sections of this paper, I will briefly present the teaching strategy and the initial part of the package which has recently been used by our research group to try a new approach in presenting historical experiments.

2. The strategy

In this teaching strategy, students' activity develops in several phases. In the first phase, students work with computer simulations of historical experiments. Essential elements of experimental devices are presented, but with limited description of their use. The students are invited to plan how to use the simulated experimental apparatuses. Of course a simulated experiment is much simpler to do than a real one but some important aspects, such as the need to plan which measurements to carry out and the difficulty of finding significant relationships among data, are unchanged.

Worksheets, designed not to guide student activity strictly but to promote reflection on the significant variables involved in the experiment and on their quantitative relations, are used by the student in this phase.

In the next phase of work, the teacher uses the information gained from the worksheets and direct observation to help students compare their interpretation of the experiments with accepted physical theory. Discussions between teacher and students should be accompanied by experiments to focus on particular aspects of the topic. It is useful for the students to consider modern versions of original experiments. This ensures that they recognize technological constraints in the original experiments and find out for themselves why certain parts of the experimental apparatus were constructed as they were in the original version. A comparison between historical and modern versions of the same apparatus can help students to understand the way the experiment was carried out and why.

Finally, students can complete their analysis of historical experiments by using the historical section of the accompanying guide which proposes, by means of a selection of original sources, a reconstruction of the different theories about the motion of falling bodies. In this way the students may see their activity of constructing and testing a model in the wider perspective of the historical construction of a coherent theory.

3. Teaching material

The list of simulations presented in our package is the following:

Part I - Galilei's experiments with an inclined plane
- A qualitative study with bells
- Quantitative study with the water clock
- Measurement of velocity

Part II - Atwood's machine
- The original Atwood's experiment
- Modern version of Atwood's experiment

Part III - Free fall
- Measurement of time by means of horizontal displacement
- A simple method to measure the acceleration of gravity
- Measurement of time with electronic devices.

Instructions for running the programs appear both on the screen and on the student guide.

In observing our students working at the computer we initially noticed that they often did not read with attention the instructions on the screen or those in their guide.

To learn from a computer program, students must become involved in activities at different perceptual and cognitive levels. In particular they must pay attention to the messages written on the screen so that they can smoothly proceed in the dialogue with the computer. This necessity seems to be in contrast with their usual attitude towards the observation of simulations.

A written message is perceived to be more appropriate for reflection: students disregard more easily a written message than a message they hear.

For this reason we have tried a new approach based on computer simulations accompanied with sound, images and animations.

In our first attempt we have used the simulation n. 1, titled "A qualitative study with bells" (Part I of the above mentioned package).

The teaching material which we have recently prepared presents first Galilei and his research work, then a reproduction of the original inclined plane existing in Florence at the "Museo di storia della scienza", and an animation which introduces the Galilei's experiment we want to study. Sound and images should, we hope, create for students both an environment familiar to them, and a richer learning context. We intend, in particular, to favour the perceptual context and to avoid interrupting the observation of computer simulations on the part of students by requiring them to read written messages.

Suggestions given by voice and images help students to use effectively the computer program which presents the simulation. 1.

Fig.1 shows the first video display referring to the Galilei's experiment described in the original manuscript known as the "bells manuscript" [2].

We tried to reproduce with the simulation the original situation in which Galilei observed the motion of a ball moving along an inclined plane, while singing a refrain to find the position of the ball at regular time intervals (he did not have a mechanical clock to measure time intervals but he, as member of a family fond of music, had a well-trained ear). So, when the student starts the ball, s/he hears a regular sequence of beeps (emitted by the computer to substitute for Galilei's refrain), while the ball is moving along the inclined plane. The student is invited to move the marks so as to obtain a "tuning" between each beep and the passage of the ball under a mark.

Then a program in which the shiftable marks behave like bells is proposed: each bell emits a beep only when the ball passes under it. The student can improve the positioning and adjust each bell until the sequence of beeps is regular.

Finally, a graduated ruler appears so that students can evaluate the displacements of the ball vs. time.

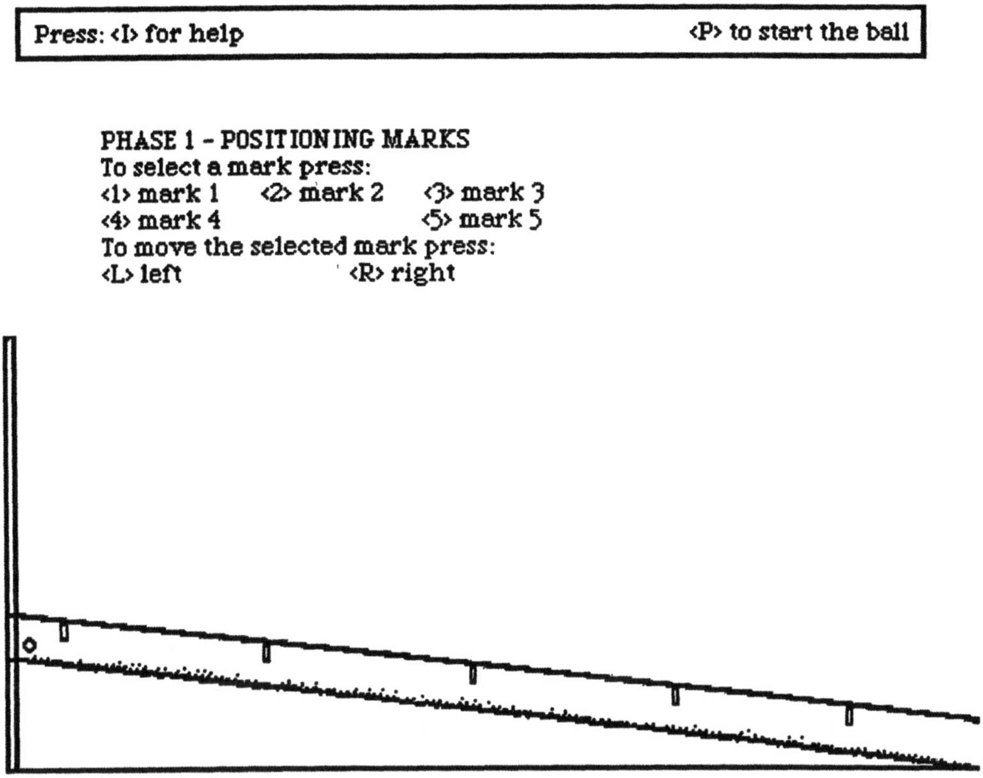

Fig. 1 Video display related to the simulation of Galilei's experiment with an inclined plane

This experiment introduces the idea that displacements along an inclined plane are proportional to the square of corresponding time intervals, and it gives students the opportunity to differentiate clearly the concepts of position, velocity and acceleration, which often, even in simple motions of objects, are not understood to be distinct.

The program allows a vivid perception of the fact that the motion of the ball is not uniform and that the velocity of the ball increases with time.

The accompanying worksheet draws students' attention to the quantitative aspect of this motion. By evaluating the displacement of the ball in successive time intervals the students can recognize that the increase of the average velocity of the ball in each interval is the same (that is the acceleration is constant).

The worksheet provided to students while using the simulation n. 1 is reproduced next.

Worksheet n. 1

Please read the displacements of the ball on the graduated ruler on the screen, and fill the table below:

Mark	1	2	3	4	5
Position					

Write the displacements of the ball in successive time intervals :

Mark	1	2	3	4	5
Displacement					

Can you obtain information about the average velocity of the ball in any one time interval?

If yes, what kind of information?

Can you infer information about the average acceleration of the ball in any one time interval?

If yes, what kind of information?

The sound is stored on a tape recorder; the program, the animation and the images on a diskette. The equipment needed is an IBM compatible computer with colour monitor and VGA adapter, and a SoundWare Drive. Even in high schools this equipment can be available and the approach we suggest could be used to present science topics at a limited cost.
The teaching material has been prepared by our research group; the mixing of sound, images, animation and computer simulations has been performed by PRISMA, a software house in Pavia[1].

4. Insights gained from pilot-testing

The initial version of our teaching material has been pilot-tested in academic years 1987, 1988 and 1989 with students enrolled in introductory physics courses at college level (a total of 189 students). The students worked in groups of two or three, sitting at a personal computer for about half an hour for each simulation. In the initial testing of the programs we noted in the students a propensity to use simulations almost as a sort of pleasant entertainment without much attention to physics concepts. As our desire was, rather, to foster a reflective attitude in the students, we responded by modifying the worksheets: instead of giving advice about the use of programs we proposed questions aimed at eliciting their reflection about the simulated experiment they were considering. (Each group was requested to fill the worksheet prepared for each experiment).

As has been mentioned, students tended to disregard the suggestions appearing on the screen and in their written guide. They were forced to re-run the program in order to fill in their worksheets. When students had problems with a computer simulation they usually sought outside help from peers, teacher and only at the end from their written guide. The discussions in each group were generally very interesting to follow as they revealed attitudes and beliefs of students. These, together with the analysis of the completed worksheets, helped us identify obstacles that seemed to hinder the development of student understanding of the kinematics concepts.

In particular, the "bells experiment" showed that:

- the students were not used to focussing their attention on the displacement of the moving ball in each successive time interval;
- it was difficult for them to determine a velocity if a measure of the elapsed time was not been made with a clock. In their answers the students correctly referred to the formal

[1] Information about the SoundWare Drive and the Authoring System used in producing the teaching material should be requested from : PRISMA 2.0 srl - Viale Cesare Battisti, 76 I-27100 PAVIA.

relation between space and time by writing v= $\Delta s/\Delta t$, but more than half of them did not know how to apply the formula in the context of the simulation. An analogous difficulty was evident in the determination of acceleration.

We found that Galilei's bells experiment is particularly effective in focussing on this problem because it was conceived precisely in order to overcome the difficulties related to the measurement of short time intervals. It is worth noting that a greater fraction of the students correctly determined the velocity and acceleration of the rolling ball in the simulation of the experiment of the water clock (Part II of the package), where they had to measure both distance and time, than in the "bells experiment".

Finally we observed that students did not read the historical guide before using computer programs, but after they completed their work with the package they showed interest in it and resorted to its use.

5. Concluding remarks

The use of computer simulation to reconstruct historical experiments seems to be promising in two respects:

- it can help to overcome the difficulty of using original apparatus;
- it can favour the understanding of the relationship between theory and experiments and the differentiation of physical quantities.

The teaching material described above could be useful to students enrolled in introductory physics courses at both the high school and college level. Teachers may plan its use in different ways, according to the class. It could, for instance, help to introduce uniformly accelerated motion after the study of uniform motion has been successfully completed.

When students have clear that the essential characteristic of a uniform motion is that in successive equal time intervals the displacements are the same, they can clearly perceive that in Galilei's experiment the situation is different.

Comparisons between the case already studied and the new one can afford students the opportunity to recognize the differences between uniformly accelerated motion and uniform motion and begin to develop quantitative understanding.

Another possible use of the material could be as a tool to reinforce the understanding of uniformly accelerated motion if its study has been completed in a traditional way (our trials of the teaching material with undergraduate students has been carried out in this situation).

Generally students are not used to considering a computer as a tool to help them think and learn. Perhaps their first experience with the computer did not orient them in this direction. It is harder for them to learn to do something which they cannot link to their other experience than something familiar.

Using computer simulations with the intention of learning from them requires the use of comprehension and study skills. Students must check whether the messages are being understood, become aware of a breakdown in their understanding, and take appropriate action to remedy it.

We believe that the approach we have recently tried helps to increase the interest of students, especially high school students and non-major students. Often, in fact, the use of computer programs requires high motivation. It can be very boring to read messages on the screen or to look for instructions on the sheets accompanying a computer program.

The recent version of our teaching material has not yet been tested: we have just completed it and we think to be able to verify its effectiveness in the next academic year. We hope that the new way of presenting the computer simulations will improve the learning environment. The multimedia approach we are suggesting should create a kind of microworld, in which students' learning is favoured. We have tried, in fact, to improve both their observation of computer simulations and their reflection on them.

Our desire is that our approach will prove to be effective and that it will serve also to show students that a computer can be an instrument that can help them deepen their conceptual understanding.

References

Bevilacqua F, Bonera G., Borghi L., De Ambrosis A., Massara C.I.: Computer simulation and historical experiments, European Journal of Physics, 11, 15-24, 1990

Drake S.: Galileo at work, The University of Chicago Press, Chicago 1978

Halloun I.A. and Hestens D.: The initial knowledge state of college physics students, Am. J. Phys. 53, 1043, 1985

McDermott L.C.: Research on conceptual understanding in mechanics, Physics Today, 24, July 1984

Viennot L.: Analysing students' reasoning: tendencies in interpretation, Am. J. Phys. 53, 5, 1985

Using computer simulations with the intention of learning from them requires the use of computation and study skills. Students must check whether the message is being understood, become aware of a breakdown in their understanding, and take appropriate action to remedy it.

We believe that the approach we have recently tried helps to increase the interest of students, especially high school students and non-science [illegible] [illegible]

[illegible]

The Design of a Learning Environment in Mechanics : Two Case Studies

Michel Caillot

Université de Picardie and LIREST, Université Paris 7, 2 place Jussieu, F-75251 Paris Cedex 05, France

Abstract: Two projects of learning environments in mechanics, still in progress, are described. Their approach is quite different. In the first project, the student has to model a physical phenomenon represented by pictures: stroboscopic photographs, series of slides or motion pictures. The software enables data acquisition and processing, and gives choice among different representation spaces. The student should build a plausible model and then confront it with data. This project is clearly based on the analysis of physicists' activity in the laboratory. The second project aims at concept learning. A major difficulty for students is the understanding of the concept of "force". In order to overcome this difficulty it seems appropriate to introduce the concept of interaction between two bodies early in the teaching. The software proposes many situations where the student has to describe which bodies interact together. Thus the student builds gradually the concept of interaction by discrimination and generalization. In the latter case, the reference is rather educational psychology.

Keywords: learning environment, physics education, modeling, concept learning

1. Introduction

Teaching mechanics to students of high school or in first year of college is a true challenge, although motion is a common phenomenon in everyday life. Newtonian mechanics is a theory, a construct which requires that some physical quantities have been introduced previously. The definitions of these quantities such as velocity, acceleration or force are rather different in their common meaning so that sudents must construct these concepts and then use them to analyze motions of real bodies. A difficulty is that when students learn mechanics, they have already developed models of phenomena which are inconsistent with Newtonian mechanics.

Microcomputers can help students to learn better the concepts of mechanics, to use their knowledge for analyzing complex situations like those found in the everyday life and confront their personal representations with scientific models. But in order to fulfill these objectives, the design of such a software should respond to a careful analysis made from two points of view: physics education and learning processes. For example the software should propose activities different of those usually found in the textbooks and enable students to have a new practice.

NATO ASI Series, Vol. F 86
Intelligent Learning Environments
and Knowledge Acquisition in Physics
Edited by A. Tiberghien and H. Mandl

In this paper we present two different approaches for learning mechanics with microcomputers. The first one is based on "practical" activities of modeling. The physical phenomena to be modeled are different kinds of motions whose the main characteristics is to be far more complex than those described in the usual textbooks. The originality of this project is to work from motions represented on pictures. The second project presented here deals with concept learning: the students have to acquire the concept of "mechanical interaction". For both projects we will show how the epistemology of physics and learning theories explain the reasons for the choices we have made.

2. Using pictures and microcomputers

2.1 Analysis of complex motions

Mechanics usually studied in high school classes applies to motions of a point mass in vacuum. They are simpler to analyze since students often do not have mathematical knowledge to study real motions in depth. But these motions are only rough approximations of motions found in the physical world: so the parabola trajectory, only valid for motion in vacuum, is an approximate model for the motion of a projectile in air.

Motions found in everyday life are far more complex than those studied in class because bodies are not point masses and friction is always present (solid or air friction). Durey [4] has shown that it is possible to analyze motions of balls thrown in some sports (tennis, golf) with the help of a microcomputer. When the modeling of such complex motion should be established, differential equations are needed which leads obviously to discarding the study in class of these motions: the students of high school (12th grade, age 18) are not familiar with differential equations and the ones in first year of college do not master them even if they have learnt how to solve them. So it is almost impossible to discuss real motions in a regular course. Another reason is that these differential equations can not be solved analytically and should be solved numerically which is impossible to do only by hand. For all these reasons, mechanics taught in class can not explain a motion so common as the motion of a soccer ball.

With microcomputers numerical calculations can be performed easily, so integrating differential equations is no longer a real difficulty. The student only has to know possible models of force law (i.e. constant friction force, air drag proportional to velocity) to model a given motion. The calculations remain completely opaque to the student. And then (s)he can confront the results of the simulation to data. If the microcomputer has solved the calculation problem, the remaining question is: "Which motions could be chosen in class to be analyzed?".

2.2 Motions represented on pictures[1]

Acquiring data from real motions is not an easy experiment to perform in class. Anybody who has tried to collect data on so common an experiment as the free fall of a steel ball, knows how such an experiment is difficult to perform even with an apparatus specifically designed to be used in physics class. In order to prevent experimental difficulties and facilitate the collection of data, specific experimental settings and special apparatus, such as air tables, have been designed. But what is the relationship between school experiments done with specific apparatus and real phenomena as those found in everyday life? With specifically designed appraratus, the distance between mechanics at school and mechanical phenomena is still increasing so that physics remains always disconnected from the real world. Even if one could expect that these experiments would work for a better student's understanding of mechanics (in fact they work badly), the student has only been able to study the mechanics of point masses in a world without friction.

Our idea is that the students collect data from real motions recorded on pictures. As they need at once the positions of an object and the corresponding times, the recording on the same picture should show position and time, in other words the motion should be decomposed into a series of snapshots. Pictures which produce such a recording are for instance stroboscopic photographs. But other ways of decomposing motion as a function of time exist: it is a motion picture where a series of successive views records the motion each 1/24 sec. or even a series of slides photographed with a motorized camera.

An attentive reader could make some criticisms concerning the use of pictures as a source of data in mechanics: Where is the experiment since physics is an experimental science ? Are you not loosing the contact with the reality that you wanted to reintroduce in your teaching ? These criticisms deserve some attention and have to be considered carefully. A first element of response is to consider the practice of the physicist or engineer in a laboratory. Pictures are sometimes source of data when other recording devices are not possible: the physicist of high energy particles uses pictures taken in a cloud chamber as data to record collisions between particles, the engineer takes pictures of car crashes against a wall to improve the security on board, the coach records a novice tennis player with a videorecorder to show later his/her poor and inefficient movements. So pictures are a way of gathering data among other souces of information. Another response is to consider that pictures are already used in physics class. All teaching of mechanics based on air tables or air pucks uses stroboscopic pictures as a starting point for introduction of different mechanical quantities (velocity, acceleration,...). And finally physics textbooks are full of stroboscopic pictures representing "real" motions such as a Big Wheel in a recreational park or a pole-vaulter in action. Often these pictures only ornament

---

[1] This project is a cooperation between the INRP (Institut National de la Recherche Pédagogique) and the University of Paris 7. It includes D. Beaufils (INRP), J.C. Le Touzé (INRP), G. Mottet (INRP), A. Texier (CNDP), M. Caillot (University of Paris 7), A. Durey (University of Paris 11) and two teacher-researchers: J. Gatecel & B. Richoux.

textbooks and do not have any instructional objective. Thus using pictures of real bodies in motion as data from which the students can apply their knowledge is a way of introducing some reality in physics classes since the purpose of physics is to understand the physical world. The students can then be more motivated to attend science classes. Once the possibility of using pictures has been established, we are going to discuss what to do with them.

2.3 Data acquisition from pictures

For data acquisition, the student uses a graphics tablet. A graphics tablet is an input device which transforms two-dimensional information displayed on a drawing or a photo into a computer readable format (see Fig.1). This device has two parts: the tablet and the cursor that is used more or less as a mouse. One clicks with the cursor on the points of interest. A low intensity signal is then emitted and the tablet picks it up, so that the cursor location is transmitted to the microcomputer under the form of two coordinates.

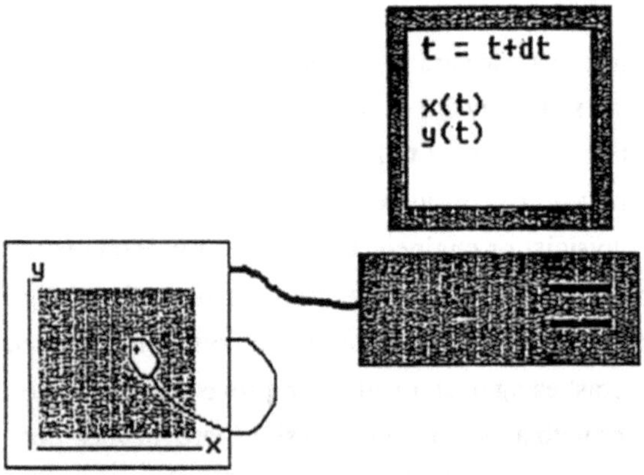

Fig.1 Graphics tablet with cursor.

A trajectory on a stroboscopic photo corresponds to a series of snapshots on the same picture. The student has to click with the cursor on the successive positions along the trajectory. If the image of the body on the photo has a large size, the student has to click on the same point to get meaningful data. But the choice of the same point all along the trajectory is not obvious for the students and corresponds already to a modeling activity (see below). With the graphics tablet, the student enters the successive positions of the body as the coordinates of a series of points. For times, either the strobe flash frequency is given, or the time separating two views of a motion picture is well defined (i.e. 1/24 sec).

Once input positions are acquired by the microcomputer, internal calculations of different kinematics and kinetics quantities are performed and are then at the student's disposal. These quantities are velocity, acceleration, kinetic and potential energies. How the calculations are performed remains completely opaque to the students.

2.4 Representation spaces

In addition of the time separating two successive positions, the set of physical quantities calculated from data includes:

- horizontal and vertical coordinates,
- horizontal and vertical velocities,
- horizontal and vertical accelerations,
- translational kinetic energy (when mass is known),
- rotational kinetic energy (when moment of inertia about an axis is known),
- potential energy,
- coordinates of the center of mass.

Once the kinematics and kinetics parameters of the motion have been calculated, they are stored in memory and then the student can display them on the screen under different formats: tables, graphs and equations.(S)he can choose a symbolic system or switch from one symbolic system to another one according to what (s)he finds the most useful for modeling the physical situation.

The software is so flexible than many types of graphs can appear on the screen. The label of two axes is completely free and depends entirely on what the student wants to model. For example the student can choose to display the horizontal velocity vs time or the vertical acceleration vs kinetic energy (a not very interesting space representation !).

In addition the software enables sophisticated calculations like smoothing curves, differentiating or integrating. For modeling, different types of parametric equations are proposed and the student's task is then to determine the parameters which fit the curves the best.

2.5 Modeling activities

The pictures used so far in an experiment with French 12th-graders (age 18) are stroboscopic motions. The next stage will be using a series of slides taken from movies representing sport situations. The experiment has been performed so far with six groups of two students coming from two different high schools. It took place after that the teaching of the regular physics course had been made.

The strobe photos given to the students represent motions such as the rebound of a golf ball, the trajectory of a shuttlecock or the motion of a wheel rolling down an inclined plane. The study of these motions is usually well beyond the syllabus of physics proposed in the French 12th-grades because these motions involve air resistance and rotating bodies. As an example, we are going to present the case of the trajectory of a shuttlecock struck by a badminton racket. We know that this motion is very dependent on air resistance because of the feathers at its end. The students familiar with motion in vacuum will be surprised when they see that they are not able to interpret their data in considering gravity as the only one force which acts on the shuttlecock.

On the photo distributed to the students only the cork is visible, the feathers are not. After that input data have been acquired by clicking on the successive positions of the shuttlecock, the student can display the trajectory on the screen, or the curves giving the coordinates vs time as well. On Fig. 2 the vertical coordinate y vs time is shown. The crosses are the recorded experimental points .

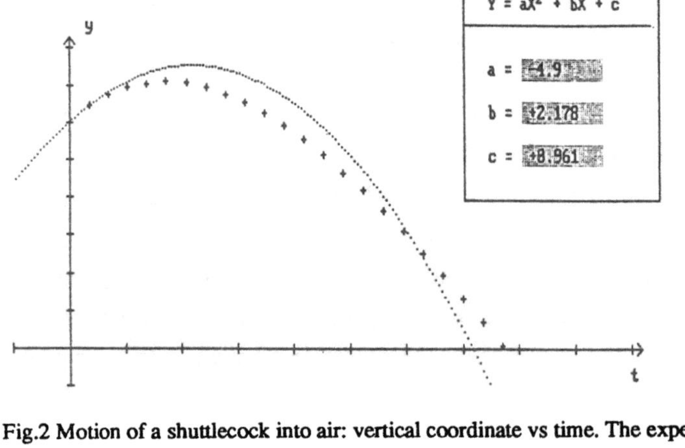

Fig.2 Motion of a shuttlecock into air: vertical coordinate vs time. The experimental points acquired through the graphics tablet are represented by crosses. The curve in dots is the parabola which results of a modeling based on a motion into vacuum.

The student knowledge is only about the parabola trajectory of a point mass in vacuum. So the first student hypothesis will be to test if the trajectory is a parabola. So for the curve y = f(t), (s)he will consider the equation

$y = (1/2)\,at^2 + V_{oy}t + y_o$

and will bind the different parameters to the values determined from other curves. So a, acceleration of gravity, is known ($a = 9.8\ ms^{-2}$), and the initial vertical velocity V_{oy} and the initial vertical coordinate y_o are determined from $y = f(t)$ and $V_y = f(t)$. Fig.2 shows the result of the simulation: clearly the model of a projectile motion in vacuum does not fit experimental data. Other ways exist to test the model of motion in vacuum. The student can also consider the curves of velocties vs time. Usually (s)he will call the graphs of horizontal (V_x) and vertical (V_y) velocities vs time, because (s)he is expecting to see a horizontal line for V_x (V_x is constant) and a oblique line for V_y (V_y decreases linearly with time), that is not the case as shown on Fig. 3.

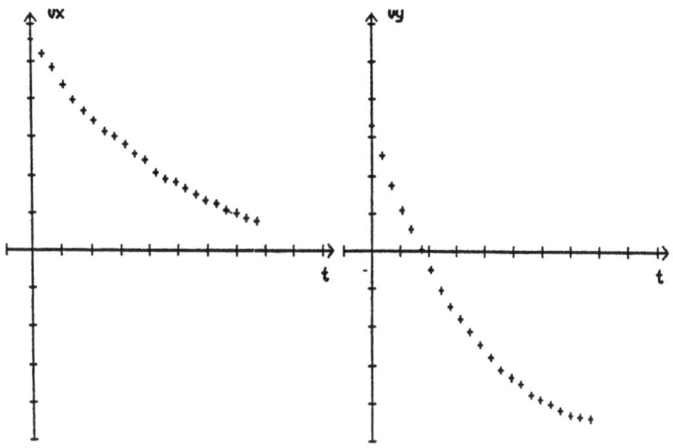

Fig.3 Motion of a shuttlecock into air: graphs of horizontal and vertical velocities vs time. With the model of projectile thrown into vacuum, the students expect to find a horizontal line for Vx and a straight line with a negative slope for Vy.

At this stage, one can say that the point mass motion in vacuum constitutes a model of zero order unable to explain the experimental curves. So the zero order model should then be refined.The student thus will try to modify the hypothesises (s)he made by taking air resistance into account. (S)he could model it either by a constant force or by a force depending on velocity (linear, quadratic or power law). Once his/her choice made, the student enters the air drag force law into the available equations and adjusts the parameters in order to get the best fit between

the experimental and the calculated curves. Fig. 4 shows the fit obtained in considering the air resistance proportional to velocity.
To do this, the student has used the microcomputer like a black box calculator which has integrated a differential equation of the second order. But the student does not know how it was performed and even (s)he is completely unable to do it analytically.

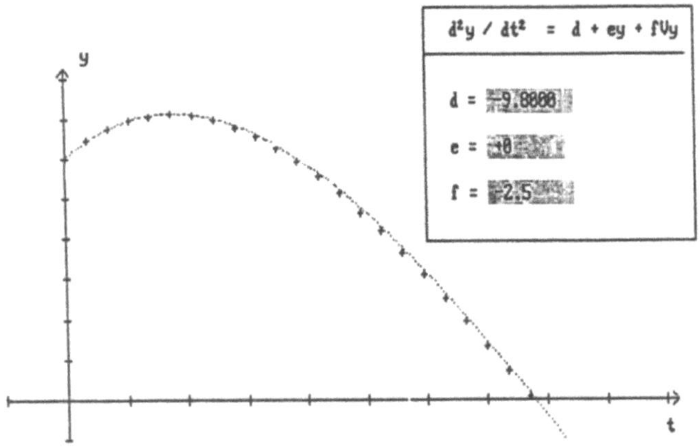

Fig. 4 Motion of a shuttlecock into air: vertical coordinate vs time. The curve in dots shows the simulation when air drag is taken into account. Air resistance is proportional to speed.

The example discussed here shows how the students are engaged in activities for modeling real situations. These activities are the usual practice of physicists doing experimental research: acquiring data, making hypotheses and testing, or using a known model and refining it after confrontation with experimental data. In this project we want therefore to promote practice in class whose reference is the physicist in his/her laboratory.

3. Concept learning

In the project so far presented, activities offered to the students come after the regular teaching of mechanics. Concepts and methods of problem solving have been taught elsewhere. In that experiment we do not know if students have understood and mastered concepts they use in their physical modeling of "reality", for example have the students' representations been changed? The question remains still open at the time where this paper is written, but the concept mastery - if it exists - can only be a by-product of modeling activitites. Conversely, software can be specifically designed for learning concepts. In the project now presented, the concept to be mastered is "physical interactions".

3.1 Concept of "physical interactions"

In the usual teaching of mechanics the key concept is "force". But this concept is often difficult to understand well. Classical errors are found in problem solving such as forgetting the contact force (i.e. normal force on a body which rests on the ground), or adding an extra force (i.e. a force associated with a kick on a soccer ball moving into air).

Previous studies of problem solving in mechanics [1, 2, 3] have shown that if the student is given opportunity to analyze interactions between the bodies present in the situation, his/her performance will be improved. When a situation is analyzed in terms of mechanical interactions no symmetry is broken: all the different bodies have the same status, none is favored and body A acts on body B as B acts on A. It is only when a system is chosen that the symmetry of interaction is broken and "force" is the concept introduced to explain how A acts on B. Before the student engages into a free body diagram on which all forces acting on body A must be drawn, we would like that students make an analysis of all interactions existing between all bodies. So the objective of software is to teach the concept of "interaction" in relationship with the nature of bodies considered.

We distinguish usually two kinds of interactions:

- contact interactions between two bodies in contact;
- long range interactions such as gravity between a body and the earth, or electromagnetic interactions.

Thus, in general, any body interacts with the earth through gravity since it owns a mass and with the other bodies with which it is in contact. However for the student the situation is much less simple than what it is stated by this last rule because several kinds of objects usually met in mechanics problems do not fulfill the condition on the mass. So strings and springs are in general modeled as being massless, pulleys as well. But there exists some cases where we have to take the pulley mass into account, as for example when the pulley turns about its axis. All objects are surrounded by air; we often do not have to consider it, excepted if it is explicitly described as a source of friction. But in this case its mass is usually discarded because it is too slight. Through these different particular cases, we see that the concept of interacting bodies depends on how the situations should be modeled.

Our project[2] is to help students to learn which kinds of interactions are relevant for a given situation. It can be used very early in the teaching of mechanics (saying for example from French 10-grades) since the concept of "interaction" is a prerequisite to the one of force. No prior knowledge is necessary to answer the questions since the objective is just learning a new concept. But it can also be used as remediation for students who have difficulties to understand the concept of "force". This project is not so well developed as the first project presented.

3.2 Learning strategy

The students learn the concept of "interaction" from examples and counter-examples. Twenty situations are presented one by one to the student. These situations are close of those encountered in usual mechanics problems. They include blocks on the ground or on an another object, pulleys, strings, springs, different charged conductors, magnets,.... They can be at rest or be in motion. Each time, the student should answer to a unique question: "Which objects interact with the object A?".

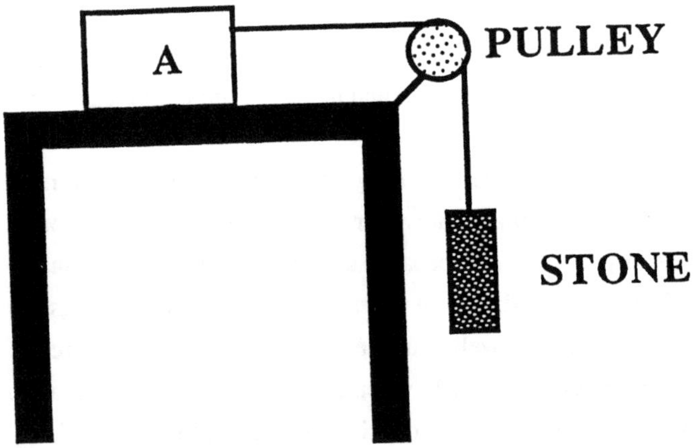

Fig. 5 Example of a physical situation on which the student should make a list of all objects interacting with body A.

---

2 This project involves M. Caillot, A. Dumas-Carré and F. Dubreuil.

Fig. 5 gives an example in which body A is in motion. Suppose that the student answers: "TABLE, STRING, PULLEY AND STONE INTERACT WITH A",

then a message will appear on the screen:

" - NEITHER PULLEY NOR STONE TOUCHES THE BODY A OR INTERACTS WITH A THROUGH A LONG RANGE INTERACTION. IN THIS CASE WE DON'T SPEAK OF INTERACTION WITH A

- YOU HAVE FORGOTTEN THE EARTH BECAUSE BOTH EARTH AND A INTERACT THROUGH A LONG RANGE INTERACTION SINCE A HAS A MASS".

Through this message the condition of contact interaction is given explicitly: both bodies must touch each other. In the same way the condition of gravity interaction is stated: it exists if the body under consideration has a mass.

If in addition the student has answered "AIR" as interacting with A, another message will appear on the screen:

"YOU ARE RIGHT TO CONSIDER AIR AS INTERACTING WITH A, BUT THE SPEED OF A IS TOO LOW TO EXPECT A MEASURABLE EFFECT ON ITS SPEED".

Again a condition on air resistance is explictly stated: air resistance will be taken into account if speed is high.

For each situation presented on the screen, thus the student learns the concept of "interaction" by induction, generalization and discrimination through examples and counter-examples. We have no ambition to assert that all concepts could be learnt like this. We are rather convinced of the contrary. Scientific concepts are not everyday life concepts gradually constructed through one's own experience. Learning scientific concepts needs to change one's representations. But in some cases like here, scientific concepts can be acquired from instances by abstracting common features, making hypotheses about the conditions of application to new instances or memorizing exemplars. The software is designed for remedial sessions when high school students or freshmen in a college are in trouble with forces and free body diagrams; it should help them to make an exhaustive analysis of all forces acting on a body or on a system since at each interaction between two bodies corresponds a force on one of both.

4. Conclusion

The two projects presented here are profoundly different. Their objectives are even contradictory. The project on using pictures and microcomputer aims to promote an activity like the physicist's. It is proposed that subsequently the regular courses of mechanics have been taught and its objectives are to engage students in activities of problem solving which can not only solved with their knowledge. The microcomputer is absolutely necessary as a calculating and graphing tool so that the students devote themselves entirely to model and simulate complex physical situations usually not studied in class. On the contrary, the project on learning the concept of a "mechanical interaction" with a microcomputer aims at the students in difficulty as remediation for a better understanding of the concept of "force". The situations presented on the screen are very similar to the ones found in any physics textbook. The program is in fact a traditional CAI program where the feedback is immediate after the student has entered his/her response. In spite of their differences these two projects have a common point: they result from research in physics education done for many years at the University of Paris 7 (and elsewhere). From our own experience, we can state that any innovation in new educational technology should be rooted firmly in research on the learning of a specific subject matter, on what we call in French "*la didactique d'une discipline* ", here "*la didactique de la physique* " (physics education).

References

1. Caillot M. & Dumas-Carré A.: PROPHY: un enseignement d'une méthodologie de résolution de problèmes de physique. In Résolution de problèmes en mathématiques et en physiques. Collection Rapports de recherches, 12, pp.199-244. Paris: INRP 1987
2. Dumas-Carré A. & Caillot M.: Résolution de problèmes et apprentissage de la physique. In G. Vergnaud, G. Brousseau and M. Hulin (Eds), Didactique et Acquisition des Connaissances Scientifiques, pp. 217-227. Grenoble : Editions de la Pensée sauvage 1987
3. Caillot M. & Dumas-Carré A.: Teaching decision making to solve textbook problems. In Mandl H., De Corte E., Bennett N. and Friedrich H.F. Learning and Instruction, Vol. 2.2, pp.67-84. Oxford: Pergamon Press 1989
4. Durey A.: Vers des activités didactiques de mise au point de modèles de physique avec des microordinateurs, [Towards didactic activities of modeling in physics with microcomputers] Doctoral dissertation , Université Paris 7, 1987

From Research in Science Education to the Conception of an ITS

M.J. Loureiro

Deptº de Didáctica e Tecnologia Educativa e Pólo de Aveiro do Projecto MINERVA, Universidade de Aveiro, 3800 Aveiro, Portugal

Abstract: In this paper some issues concerning computers in physics teaching are presented. It is supported that the teaching/learning process, and therefore an ITS, must be seen as a knowledge interaction system. It is claimed that if one wants to help the students to modify their ideas one ought to develop ITS starting from those ideas. A teaching sequence based on a constructivist learning theory is presented, with the results of its application. The strategies used are useful for ideas that can be tested at a macroscopic level but less efficient for more abstract ideas, their efficiency depending on the topic under study, and on the pupil. In the teaching /learning model of an ITS one ought to describe the most useful strategies to go beyond a problem.

Keywords: computer assisted learning, intelligent tutoring systems, science education, alternative conceptions, teaching strategies, electricity

1. Introduction

There have been many researchers over the past twenty years who have worked in the field of the student's alternative conceptions [11, 12, 22]. Recent research has focused on teaching strategies taking into account those alternative conceptions [13, 17, 18, 25, 26]. Moreover, research on Artificial Intelligence has provided some useful techniques for educational purposes [24, 30]. However the majority of the implemented Intelligent Tutoring Systems (ITS) do not take into account the work carried out in Science Education.

This article presents a research project being developed at the University of Aveiro, Portugal, which has attempted to integrate studies and research in Science Education with studies in the field of ITS. In order to do so, the presentation was divided in three items: i) the project objective; ii) some issues concerning computers in Physics teaching; iii) description of the project (phases, teaching sequence and strategies, and preliminary results).

2. Project objective

The main objective proposed was the construction of a program able to adapt itself to each students' particularities. Such adaptation depends on two factors [9]: number of variables concerning the student's cognitive and affective structure considered by the system, and the pertinence of those variables.

Because of the diversity of the variables concerning the student's cognitive and affective structure, it becomes rather difficult to build a system considering all of them together. In this way, it was decided to give special relevance to one of those variables, that is the student's alternative conceptions, those ideas that the students develop before or during formal education.

The decision named above was made considering that the main source of information of the program are the "errors" made by the student [10], and such "errors" have been explored in a rather superficial way. It has also been considered the present evidence about the origin of those "errors", the alternative conceptions, which may be very resistant to teaching strategies that simply ignore them.

Therefore, in summary, the main program objective is to help the student to become aware of their alternative conceptions and eventually modify them.

3. Some issues concerning computers in Physics teaching

There are three types of computer's programs on Physics teaching: i) computer assisted instruction (CAI); ii) computer assisted learning (CAL); iii) computer based laboratory (CBL) [19, 21].

CBL programs are the best to facilitate the acquisition and treatment of laboratory data. General purpose programs and exploration environments, like simulations and Hypercard programs, are examples of CAL programs. Tutorials, and drill and practice programs can all be classified as CAI programs.

Figure 1 shown the differences between CAI, CBL, and CAL programs concerning the student's initiative and guidance. By using CAL and CBL the students can do whatever they want. However by using CAI the student have to follow the program's decision, because the teaching situations are defined from the beginning. The inverse process can be verified in relation to the student's guidance. Thus, both approaches may offer some disadvantages. CAI programs do not allow the student to use his curiosity. By manipulating CAL or CBL programs, the student can get stuck or use the time inefficiently [3, 4].

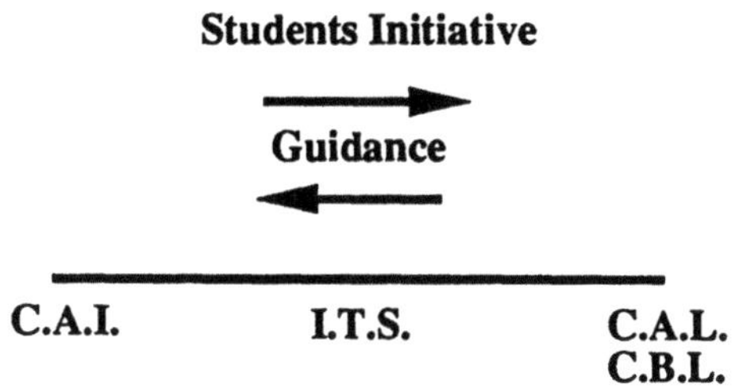

Fig. 1. Student's initiative and guidance in CAI, CAL and CBL programs and ITS.

Given the objective already described, the aim is to develop a program that allows the student some initiative, and, at the same time, interrupt him when necessary, to give him some guidance or explanations. It was decided that this could be done by using Artificial Intelligence techniques. These type of programs lies in the middle of the continuous represented in Fig. 1.

3.1 Subjacent learning theories

The CAI programs start from the assumption that learning is performed in a monotonic way, that is, provided that the student already have the necessary pre-requisites, then a new skill can be learned. The student is expected to absorb knowledge in an incremental fashion and is viewed as a blank slate or as an empty glass [27]. In this approach the program is seen as a Knowledge Communication System (KCS), with the program being the source of knowledge. The ITS programs derived from expert systems can also be considered as KCS [7], where the the source of knowledge is the expert model.

A constructivist learning theory supports CAL and CBL programs. The teaching/learning process is centered in the construction of knowledge by the student himself. Moreover, it is considered that all the students, starting formal education, bring some knowledge to the task of learning a new skill.

We claim, as other defend [7], that the teaching/learning process must also be seen as an interaction between the program/teacher and the student, and between students. All those involved in the process learn and have some knowledge at the beginning. Thus, we need to develop Knowledge Interaction Systems, programs with learning abilities.

3.2 ITS architecture and some research problems

Theoretically ITS has five blocks: the interface; the student model; the teaching/learning model; the expert model; and, the control program (the so called "tutor") [24].

The rules defining the domain to be learned are represented in the expert model. This model in almost all of the ITS that has been constructed is defined with rules and knowledge derived from interviews with experts. The main concern here is the fact that this knowledge may not necessarily be the one that can be taught, in particular at the initial level. The problem here concerns the didactic transposition and questions like what type of concepts, principles, models, capabilities, can be learned by the student at a certain level and how to analyse the domain to be learned, in summary, how to adapt the domain to the student.

Other questions have been raised which may have some implications in the development of the expert system. One concerns the types of knowledge which should be represented [1]: procedural, declarative, or causal/qualitative [31]. Taking into account the different tools used for prediction at different levels, for instance, natural language, qualitative relations, quantitative relations and theories [28], maybe one should no longer speak about expert systems, but instead about the knowledge base of the system, and build it using multiple representations.

The student model has information concerning the cognitive and/or affective structure, that characterize the student. This information permit the ITS to adapt the teaching/learning process to the student's particularities. Many research problems were pointed out during the workshop [20, 23] concerning the construction of the student model. The lack of information, among Artificial Intelligence researchers, about what the alternative conceptions held by students in particular domains are, is another one. In relation to Physics concepts very few systems consider this type of knowledge. The work being done in Science Education should be the base for the inplementation of this model.

The teaching/learning model, has a learning theory behind, and is constructed from rules where the conditions are relied upon, for instance, the diagnostic of the student's cognitive state, and his learning style. The actions to be performed are the strategies that, in principle, are better suited to help the student to progress, or to modify the state of the student.

According to some authors [10, 15] the majority of the researchers, working in ITS, are not explicit in relation to the pedagogical strategies implemented in their systems. Another complaint concerns the lack of didactic principles that will permit to determine the strategies that should be adopted in general and also in particular domain/situation.

Research should be done to define and test teaching strategies to modify the students' alternative conceptions. Reactions to those strategies should also be identified in order to do the student modelling.

4. Description of the project

We have defined the following phases for the project: i) a study of the alternative conceptions in electricity held by 12 to 17 years old pupils. This study was concluded at the beginning of 1988; ii) the definition of the subject topics and the program's objectives; iii) the definition of the teaching sequence, teaching situations and strategies to help students to modify their alternative conceptions; iv) implementation of the program.

The third phase mentioned above has been done from March to June 1990 and will be presented as well as some preliminary results of its application.

4.1 Teaching sequence and strategies

Concerning the cognitive domain the program's objective will be to help students to make predictions in simple electrical circuits.

In order to structure the topics it was made a bibliographic review about the teaching sequences preconized in some Portuguese text books as well as in the literature.

The analysis of the Portuguese text books [see, for instance, 8] has led to the conclusion that the electric circuit is, frequently, introduced from a linear sequence of terms, definitions, and rules following the hierarchy Q -> I -> V -> R (where Q = electric charge, I = electric current, V = potential difference and R = electric resistance). This sequence is problematic at different levels. Among them, as similarly reported by Hartley [16] concerning German text books, they do not put in evidence the electric circuit as a system that has a certain complexity and interdependent relations between the subjacent concepts; and, the strategies used are too operational, they get the students to acquire rules to quantitatively analyse the circuits without understanding the underlying concepts.

From our point of view the basic electricity concepts must be introduced at the same time as the electric circuit. The electric circuit has to be considered as a causal system where there are energy transference, or with the role of transferring energy. Thus, the first conceptual term used is energy.

There is a big discussion about; i) at what level of studies should we introduce energy [29]; ii) and, whether we must use the concept of energy transference or energy transformations [14]. Even knowing that the concept of energy is a very abstract one, we have decided to use it because, in Portugal, energy is a term very often used by the media, therefore familiar to the student, and is introduced, in school, three years before the study of electricity. In other hand the properties attributed by the students to electric current are similar to those scientifically associated with energy. Yet another reason is related with the conviction that to motivate students to learn about electric circuits we must show them their utility.

The main questions to be raised by the teaching sequence should then be: question 1 - what is necessary in order to have energy transference in an electric circuit? Answering this make it possible the familiarization with the elements of a simple electric circuit and the closed circuit rule; question 2 - why and how does it happen? This question lead to the concepts of potential difference and electric current, and to the association of electric charge with electric energy carriers; question 3 - how can we change it? To answer this question the association of sources, the electric resistance and associations of resistors must be discussed.

From March to June 1990 the teaching sequence presented in table 1 was followed by 20 pupils. Pupils were at the eight level of school, which is the first year of Physics teaching. The average age of those pupils was 16 years, because they were all repeating the school year. For that reason, to this class special teaching strategies have been used.

At the beginning and at the end of the teaching sequence the same diagnostic test was applied. The objective of that test was to diagnose the pupils' alternative conceptions about the concepts to be learned and to find out the effects of the approach used.

Over the last years, some teaching strategies have been preconized as being sequences that facilitate the modification of student's alternative conceptions, helping them to learn the scientific ones [6]. All those strategies are inscribed in a constructivist view of learning. One of their essential characteristic is to involve students. Those sequences give a lot of relevance to the diagnostic of the students' alternative conceptions and to the cognitive conflict phase. Nevertheless, we think that, the cognitive restructuration and the application of the new conceptions are also fundamental.

For each of the topics presented in the teaching sequence the teaching strategy used is the one outlined in Figure 2.

The aim of the diagnostic phase was to familiarize students with the electric circuits and to give students the possibility to express their views. In this phase we have used, mainly, sheets with tasks. For instance, in order to diagnose the students' ideas about how to change the potential difference, the task was to predict (justifying) the potential difference of a battery, two serial batteries and two batteries in parallel.

Table 1 - Structure of the teaching sequence

Teaching time	Essential topics
2 hours	Motivation for the study of Physics. Physics: its object and methods. The interface between Physics, Society and Technology
1 hour	Forms of energy and their manifestations.
2 hours	Motivation for the study of Electricity. Production and transference of electric energy. Sources and receptors. Dangers of the Electricity.
2 hours	Manipulation of simple electric circuits (a bulb and a battery). Polarity of the elements. Closed circuit rule.
8 hours	Sources, the Volta battery. Potential difference, the volt and the voltmeter. Associations of sources. Electric current, its direction and conservation.
6 hours	Electrical conductors, electric resistance and associations of electric resistances.
3 hours	Relations between the variables, potential difference, electric current, electric resistance, and electric energy.

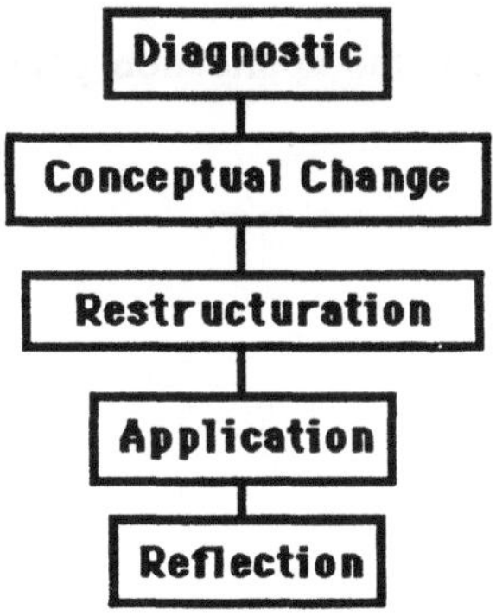

Fig. 2. Teaching strategy

The conceptual change phase is a crucial one. We have tried different activities to put the students in cognitive conflict and help them to change their conceptions. Small (2-3) work groups doing experimental work were used, discussions with the peers and the teacher were encouraged, daily-life observations were recalled, we have also utilized some particular objects or situation. An example of a particular object used was a crashed bulb, so students could see that the bulb has two poles.

The restructuration phase consisted of teacher-led discussion, focused on the meaning of the results of the experimental work, and of the relations among the observations. Conceptual networks were filled or made, at that stage, to put in evidence the relations between the concepts. Figure 3 represents the conceptual network build to relate the main electricity concepts.

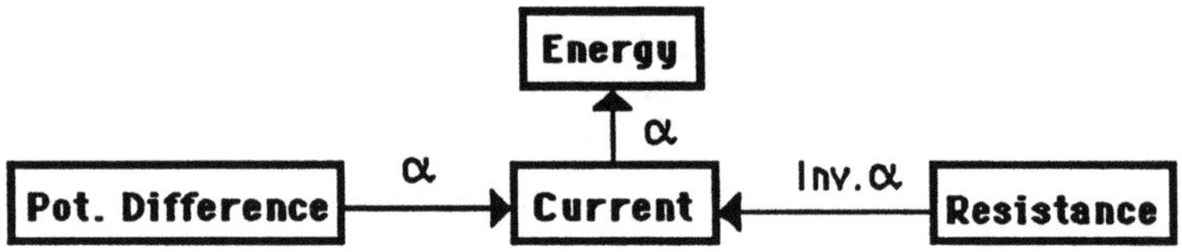

Fig. 3. Relations between the main electricity concepts

The application phase was meant to facilitate the transference of the new conceptions to daily-life situations. Figure 4 shows one of those situations for the application phase concerning the first question.

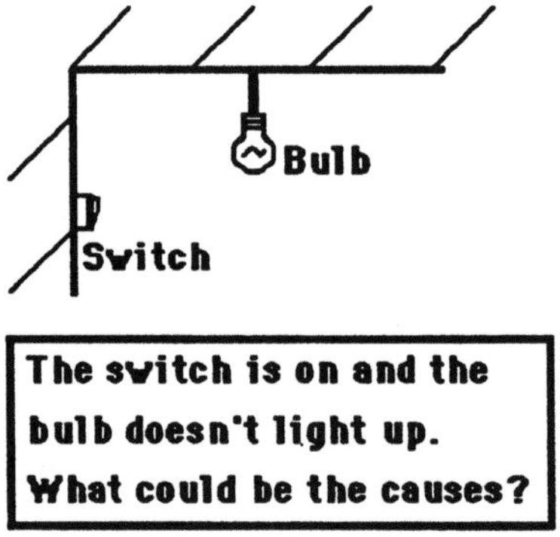

Fig. 4. Example of a daily-life situation used in an application phase.

In the last phase, the reflection phase, the pupils were asked if their conceptions had changed and which activity helped them most. One of the objective was to aware them for the conceptual change. The results of this phase will be used in the development of the teaching/learning model, the program will propose first the activities that the students of the experimental study thought helped them most.

4.2 Preliminary results

In this section it is presented the preliminary results of the experimental study for the polarity of electric circuits elements, the closed circuit rule and the direction and conservation of the electric current (table 2). Those results are preliminary because it has been analysed only the diagnostic and post-test results, and the results obtained in the reflection phase.

In the right column of the table 2 we present the frequencies for less clear justifications, absent justifications and absent answers.

From the choices made by pupils in the question Q1.1 we have inferred their ideas about the polarity of the bulb and the battery. The justifications given reflect the pupils' ideas about the cause that makes the bulb light up and therefore are not presented here.

The frequencies of the scientific answer (S.A.) and of the classified justifications to the questions Q1.3, Q2.2, and Q2.4 have been calculated. The objective of the question Q1.3 is to test ideas about the closed circuit rule. Questions Q2.2, and Q2.4, are related with the conservation and the direction of the electric current.

For the question Q2.1, where it was tested if the pupils thought that the current flows all over the circuit, to see how they conceptualize the closed circuit rule, we present the frequencies of all the choices. The frequencies concerning the justification are also presented.

Figure 5 represents a simplified version of the questions used to diagnose pupils' views concerning the topics named above.

4.2.1 Results of the Questions Q1.1, and Q1.3

From the choices made in question Q1.1 we can say that, before the teaching sequence, the majority of the pupils knew that the battery is a bipolar device and were not aware that the bulb has also two poles (pupils chose the pictures "b", "d", and "f" - Elem. U./Bip). After the teaching sequence, only 2 answers can be classified in that category, all the other answers were that the bulb will light up only in the picture "d" (Elem. Bip).

Table 2 - Diagnostic and post-test results (N=20)

Quest. /Test		Classified choices						N.clear Don't Know
		Elem. Bip	Elem. U./Bip.					
1.1	D.	4	16					
	P.	18	2					
		S.A.	Bulb-B. or path	Path.	Circ.(En. /E.C/El.)	Lighted bulb		
1.3	D.	6	1	2	9	3		5
	P.	16	4	4	9	1		2
				Wire-justifications				
		S.A.	Choice1	EC driven	Electron pass	Energy transfer.	To light the bulb	
2.1	D.	2	12	4	1	0	3	
	P.	8	4	5	5	2	1	
			Bulb-justifications				B. Just.	
		Choice2	Bulb light	Electron EC pass	Energy transfer.	Choice3	Batt. Delivers	
2.1	D.	3	1	0	0	3	2	11
	P.	6	2	2	4	2	4	2
		S.A.	EC flows	EC used	Part EC used	EC transfer.	EC light bulb	
2.2	D.	3	2	4	10	1	8	7
	P.	14	5	2	4	6	1	8
		S.A.	Name	EC unidir.	EC bidir.			
2.4	D.	10	1	1	9			9
	P.	19	9	7	1			2

The diagnostic test results of question Q1.1 are not surprising since all the pupils were attending Electricity teaching, for more than one year. The manipulations usually done, when pupils have the opportunity to manipulate simple electric circuit, use batteries and bulb in sockets. From those manipulations pupils had learn, in previous years, that the battery is bipolar but they did not become aware that the bulb is bipolar too.

Pupils choices for question Q1.3 reflect that, after teaching, 16 of pupils considered that an electric circuit is closed in the pictures where a short-circuit is represented, and in the picture were the bulb lights up (choices "b", "f" and "d"). The justifications for that question were classified into four categories. Two of them had a technical character, namely the choices: -Bulb-B. or path- (where the picture with a lighted bulb was differentiated from the short-circuit), and -Path- (where that differentiation was not done). The justification -Circ. (En./E.C./El.)- (something, energy, electric current, or electrons, flows) is conceptual. The justification -Lighted bulb- is causal and utilitarian.

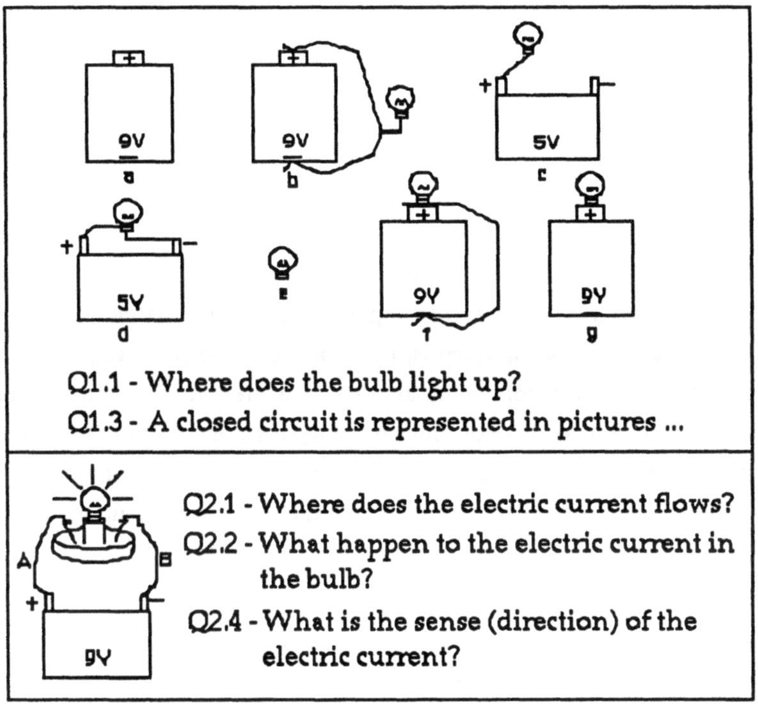

Fig. 5. Simplified version of the questions used.

4.2.2 Results of the question Q2.1

Pupils' replies to question Q2.1 show that, before the teaching sequence, 12 pupils thought that the electric current, in a closed circuit, flows only in the wires (choice 1). After instruction, 8 students considered that the electric current flows all over the circuit (S.A.), 6 took into account the wires, and the bulb (choice 2), and 2 the wires, and the battery (choice 3).

The justifications had elements to justify why the electric current flows in the: wires (wire-justifications); bulb (bulb-justifications); and, battery (battery-justifications). There are conceptual justification, like, for instance, electric current is driven in the wires (E.C. driven), electrons pass in the wires (or in the bulb), or energy is transferred in the wires, and electric energy is transformed in the bulb. There are also utilitarian justification, as electric current flows: in the wires to light up the bulb; in the bulb because it is lighted up; in the battery because it is the battery that delivers the electric current. The majority of the justifications, except for the battery, were of the conceptual type.

From those results it cannot be considered that the activities used related with that topic were efficient. This maybe due to the fact that we cannot do any experiments to show pupils that the electric current flows in the bulb, and in the battery. We think that a good computer simulation could help pupils to clarify that view.

Some researchers [2, 5] have tried to use analogies to explain what happens at the microscopic level. We have avoided that because: i) analogies must be used very carefully, we cannot transfer all the characteristic between systems; ii) the use of an analogy requires the ability to transfer knowledge from one context to another. Transference is a high level capability that we want the student to develop during a Physics course, namely in the application phase of the concepts. The question is, how can we use as a pre-requisite a capability that we want to develop?; iii) we had constraint with time: all the teaching sequence had to be pursuit in about 28 hours.

4.2.3 Results of the question Q2.2, and Q2.4

The results of question Q2.2 show that, after the teaching sequence, 14 pupils chose the scientific answer - E.C. flows without being used. 6 pupils still chose the item that stated that the electric current is all used, or in part, in the bulb. Those results are acceptable though better results were obtained by others [17, 18, 25].

Concerning the justifications we cannot be so confident. 5 students explained why the electric current flows around the circuit without being used but 6 claimed that the electric current flows, and is transformed (EC transfor) in the bulb. This last statement show that those students had problems to discriminate the concepts electric current, and energy. Perhaps we have to review the unit dedicated to the discrimination of the concepts.

In relation to the direction (Q2.4), before the teaching sequence, 10 pupils knew already that the electric current has only one direction (E.C. unidir.) but they could not justify. After the classes, 19 chose the scientific answer but only 7 justified it correctly. Three types of justifications where identified: i) some pupils merely expressed that it is like that because it is the real or the conventional sense (direction) of the electric current (Name); ii) the justification -EC unidir.- is the scientific one (electric current flows from the negative to the positive pole of the battery because the negative pole has an excess of electrons, and the positive a deficiency); iii) pupils who had a bidirectional view of the electric current flows (EC bidir.) stated that, for instance, there are charges going from the positive, and the negative, poles of the battery to the bulb, to light it up.

4.2.4 Results of the reflection phase

Table 3 presents the results of the reflection phase in relation to the topics described above. The first range presents the results concerning the polarity of the electric circuits elements, and the closed circuit rule, and the second those related with the electric current. The choices where A=manipulation of electric circuit, B=teacher explanations, C=discussions with peers, D=research in books, encyclopedias, ..., and E=particular object or situation.

As it can be observed, from table 3, more students thought they had changed their view in relation to the topics of electric current than for the polarity of the elements, and the closed circuit rule (column Conc. change). This result is consistent with the foregoing, since, at the beginning of the teaching sequence a larger number of pupils had a scientific idea about the polarity of the elements, and the closed circuit rule, than for the electric current (except for the direction of the electric current).

Pupils opinion about what strategy helped them most are not the same, and vary with the topics under study. Table 3 shows the frequency for the first choice (column Why? 1th Ch.), and if a given strategy correspond to the first or second choice (column Why? 1th/2th Ch.).

Those results led us to the conclusions that, for the pupils with whom we have worked, the most useful strategies are: i) discussions with peers followed by manipulations of simple electric circuits (trying to light up a bulb), and the observation of broken bulb (particular object), for the polarity of elements, and the closed circuit rule; ii) manipulations of electric circuits (the ammeters experiment, that consists of having a simple battery-bulb circuit with an ammeter before the bulb, and another after), and teacher explanations, for the electric current.

Table 3 - Results of the reflection phase.

Topics	Conc. change	Why? (1th. Ch.)					Why? (1th/2th. Ch.)				
		A	B	C	D	E	A	B	C	D	E
Closed circuit and pol. of elements	12	3	3	5	0	4	7	6	10	0	7
Electric current	17	10	7	0	0	0	17	16	0	1	0

5. Conclusions and implications of the preliminary results

In general, from the preliminary results, it can be said that the results obtained, for the multiple choice are better than those for the justifications. This may be due either to deficiencies in conceptualizing, held by the pupils, or derive from their poor abilities to write, and express themselves. The strategies used to modify the ideas that can be tested at the macroscopic level were efficient. The same cannot be claimed for the more abstract conceptions that require discrimination of concepts. It can also be said that the results presented here were not so good as the results reported by others [17, 18, 25]. This implies that part of the teaching activities must be reviewed, but it cannot be forgotten that the class was a special one (see above).

The variation with the topics under study, and from pupils to pupils of the preferred strategies used supported some previous ideas. The first one was that there are not good, or bad, strategies, their use must be relative to the topics in study, and vary with the students.

As can be derived from the description of the project's phases, we have to face now the implementation of the program. One can ask how are the results of the previous work going to be used?

The student model will be constructed from the results of the alternative conceptions study and the results of the teaching sequence, strategies, and activities, experimentation. Since the alternative conceptions held by the students depends on their age, school level, history of interaction with the program and objectives to pursuit, it can be said that to develop that model one must construct more than a library of alternative conceptions. Some heuristics describing the alternative conception to be tested, given the student's characteristics, ought to be implemented.

The teaching sequence, situations, and the strategies defined, plus the results of their experimentation, will be the base for the development of the teaching/learning model. A direct consequence of the preliminary results presented, namely those of the reflection phase, is that if one want to individualize instruction, and simulate interactions between the pupil, and the teacher, one must look for the causes of pupils "errors" not only in the student model but also in the teaching/learning model. In the teaching/learning model we ought to describe for each alternative conception, what are the most useful strategies to help the student to overcome it. This means that for one condition (state of the student) more than one action must be designed.

In order to make this model self-improving, it would be interesting to define some rules to change the priorities of the actions according to the results of the program manipulation by the students.

Acknowledgement

We would like to thank Mº Celeste Caetano and José Manuel Lopes for their support in the development, and experimentation, of the teaching sequence and strategies, and Profs. Nilza Costa and Rosangela Bertelli for the revision of that paper.

References

1. Anderson, J.R.: The Expert Module. In: Foundations of Intelligent Tutoring Systems (Polson, M.C. and Richardson, J.J. Eds.). London: Lawrence Erlbaum 1988
2. Black, D. and Solomon, J.: Can Pupils use Taught Analogies for Electric Current. School Science Review. Dec.-87, 249-254 (1987)
3. Cabrol, D.: Le Rôle de l'I. A. dans l'Enseignement de la Chymie. To be published in: Proc. of the "1º Encontro Computadores no Ensino da Física e Química". Universidade de Coimbra, Coimbra. 24-28 Fev. 1990
4. Clancey, W.J.: Intelligent Tutoring Systems: A Tutorial Survey. In: Current Issues in Expert Systems (van Lansweerde, A. and Dufour, P., Eds.). London: Academic Press 1987
5. Closset, J-L.: D'où Proviennent Certains "Erreurs" Rencontrés chez les Elèves et les Etudiants en Electrocinétique? Peut-on y remédier? Bulletin de l'Union des Physiciens. 657, 81-102 (1983)
6. Cosgrove, M. and Osborne, R.: Lesson Frameworks for Changing Children's Ideas. In: The Implications of Children's Science (Osborne, R. e Freyberg, P., Eds.). London: Heineman 1985
7. Costa,E.: Present and Future of Intelligent Toturing Systems. To be published in: New directions for Educational Technology (O'Shea, T. , Ed.) . N.Y.: Springer-Verlag
8. Cruz, M. N., Martins, I.P. and Martins, A.: À Descoberta da Física - 8º Ano de Escolaridade. Porto: Porto Editora 1985
9. Depover, C.: L'Ordinateur Média d'Enseignement: un Cadre Conceptuel. Bruxels: De Boeck-Wesmael 1987
10. Dillenbourg, P.: Conception et Réalisation d'un Tutoriel Evolutif Exploitant les Techniques de l'Intelligence Artificielle. Université de Mons, Mons 1986
11. Driver, R., Guesne, E. and Tiberghien, A. (Eds.): Children's Ideas in Science. Philadelphia: University Press 1985

12. Duit, R., Jung, W. and von Rhoneck, C. (Eds.): Aspects of Understanding Electricity, IPN-Arbeitsbericht 59, Schmidt und Klauning, Kiel 1985
13. Dupin, J.J. and Joshua, S.: Experimentation d'Approches Hypothetico-deductives de la Physique en Classe de Seconde: Conditions et Evaluation (Vol. 1). Marseille: Groupe de Recherche en Didactique de la Physique 1989
14. Ellse, M.: Transferring not Transforming Energy. School Science Review. March-88, 427-437 (1988)
15. Halff, H.M.: Curriculum and Instruction in Automated Tutors. In: Foundations of Intelligent Tutoring Systems (Polson, M.C. and Richardson, J.J. Eds.). London: Lawrence Erlbaum 1988
16. Hartley, H.: The Electric Circuit as a System: A New Approach. European Journal of Science Education. 4(1), 45-55 (1982)
17. Licht, P.: A Strategy to deal with Conceptual and Reasoning Problems in Introductory Electricity. In: Proc. of the Second International Seminar in Misconceptions and Educational Strategies in Science and Mathematics. Cornell University, Ithaca, N.Y., July 26-29, 1987
18. Licht, P.: Diagnosis of Pupil's Preconceptions with a Multiple Choise Test and the Effects of its use in a Teaching Strategy. In: Proc. of the 14th Annual Conference of the ATEE. Kristianstad, Sweden, Sept. 3-8, 1989
19. Loureiro, M.J., Vasconcelos ,N., Caetano, C. and Lopes, J.M.: Computadores no Ensino da Física e Química: Análise Multidimensional. To be published in: Proc. of the "1º Encontro Computadores no Ensino da Física e Química". Universidade de Coimbra, Coimbra. 24-28 Fev., 1990
20. Mandl, H.: Diagnosis System for a Learning Environement on Physical Superposition of Motion. To be published in: Knowledge Acquisition in the Domain of Physics and Intelligent Learning Environement (Mandl, H. and Tiberghien, A., Eds.). N.Y.: Springer-Verlag
21. O.C.D.E.: Technologies de l'Information et Apprentissage de Base: Lecture, Ecriture, Sciences et Mathématique. Paris: O.C.D.E. 1987
22. Osborne, R. e Freyberg, P. (Eds.): The Implications of Children's Science. London: Heineman 1985
23. O'Shea, T.: Modelling Physics Problem Solving. To be published in: Knowledge Acquisition in the Domain of Physics and Intelligent Learning Environement (Mandl, H. and Tiberghien, A., Eds.). N.Y.: Springer-Verlag
24. Polson, M.C. and Richardson, J.J. (Eds.): Foundations of Intelligent Tutoring Systems. London: Lawrence Erlbaum 1988
25. Psillos, D., Koumaras, P. and Valassiades, O.: Pupils' Representations of Electric Current Before, During and After Instruction on DC Circuits. Research on Science and Technological Education. 5(2), 185-199 (1987)
26. Shipstone, D.: Pupils' Understanding of Simple Electrical Circuits: Some Implications for Instruction. Physics Education. 23, 92-96 (1988)
27. Those pictures were presented by V. Teodoro during a discussion in the "1º Encontro Computadores no Ensino da Física e Química". Universidade de Coimbra, Coimbra. 24-28 Fev., 1990
28. Tiberghien, A.: Introduction to an Epistemological Analysis of Modelling in Physics from a Didactical Point of View. To be published in: Knowledge Acquisition in the Domain of Physics and Intelligent Learning Environement (Mandl, H. and Tiberghien, A., Eds.). N.Y.: Springer-Verlag
29. Warren, J.W.: At What stage Should Energy be Taught? Physics Education. 21, 154-156 (1986)
30. Wenger, E.: Artificial Intelligence and Tutoring Systems. California: Morgan Kaufman 1987
31. White, B. and Frederikson, J.: Qualitative Models and Intelligent Learning Environments. In: AI and Education (Lawler, R. and Yazdani, M., Eds.). N.J.: Ablex Publishing Corporation 1987

Research as a Guide for the Design of Intelligent Learning Environments[1]

Lillian C. McDermott

Department of Physics FM-15, University of Washington, Seattle, WA 98195, USA

Abstract: This paper discusses how research on the teaching and learning of physics can guide the design of intelligent learning environments. The example used to illustrate the process is a computer program that is intended to help students learn how to relate a motion to its graphical representations. The discussion is divided into two main sections. The first of these describes how research helped improve the effectiveness of the program as a teaching aid. The second part demonstrates how the computer program was used as an investigatory probe to examine student understanding of motion graphs. The paper concludes with a commentary about the importance of conducting research throughout the development of computer-based instructional materials.

Keywords: computer, education research, student understanding, kinematics, motion graphs, instructional aid, research tool

1. Introduction

When the computer was introduced into physics education, many physicists were enthusiastic about its potential for improving instruction. Advances in technology gradually made possible the creation of an array of new learning environments, such as simulations, microworlds, and real-time graphical representations. It became feasible to try to tutor, monitor, and test students on an individual basis. Over the years instructors, curriculum developers, and programmers have produced vast quantities of software. Nevertheless, although there has been much progress, we find today that the computer has made little difference in physics education, except perhaps through use as a laboratory tool.

[1] The work reported in this paper has been supported in part by the National Science Foundation under grant MDR 84-70081.

NATO ASI Series, Vol. F 86
Intelligent Learning Environments
and Knowledge Acquisition in Physics
Edited by A. Tiberghien and H. Mandl

Our experience in the Physics Education Group at the University of Washington suggests that a major reason why computer-based instruction has not been very effective is its failure to match the needs and abilities of students. This paper illustrates how research can help improve this match.[1] The discussion is in the context of a computer program, *Graphs and Tracks,* by David Trowbridge[2]. The purpose of the program is to help students learn how to relate an actual motion to graphs of its position, velocity, and acceleration versus time.

The paper is divided into two main sections. In the first of these, we can think of research as driving the computer program. We describe how research provided the initial motivation and guided the later development of *Graphs and Tracks*. This discussion is followed by a second shorter section, in which we discuss how the same program was used to investigate student understanding of motion graphs. In this case, the program itself determined the direction of the research.

2. Research drives computer program

Development of *Graphs and Tracks* extended over several years. The idea of basing instructional software on a physical system consisting of a ball and tracks evolved from Trowbridge's research and teaching experience while a doctoral student with the Physics Education Group. The topic for his dissertation was an investigation of student understanding of the concepts of velocity and acceleration in one dimension.[6] Most of the data in this research were obtained during individual demonstration interviews in which students compared the motion of two balls rolling along adjacent tracks.

While the investigation of conceptual understanding in kinematics was still under way, the Physics Education Group was also beginning research on difficulties that students encounter in drawing and interpreting motion graphs.[2] Part of this effort involved the design of laboratory-based materials to address some of the specific difficulties that were identified.[3] The resultant curriculum was incorporated in an instructional module entitled *Kinematics*.[4]

The present version of *Graphs and Tracks* has two parts: Part I (From Graphs to Motion) and Part II (From Motion to Graphs). Both parts involve the motion of a ball rolling on a set of tracks that can be configured in different ways.

---

[2] This program won the 1988 NCRIPTAL award for best physics and best integrated software. NCRIPTAL is an acronym for the National Center for Research to Improve Postsecondary Teaching and Learning. The center administers a higher education software awards program for EDUCOM, a consortium of 500 colleges and universities dedicated to information technology in higher education.

2.1 Development of Part I

After leaving the University of Washington, Trowbridge went to the University of California at Irvine, where he began trying to implement on the computer some laboratory-based activities in *Kinematics*. He focused his attention on a set of experiments designed to help students relate the graph of a motion to the actual motion of a ball on a set of tracks. Preliminary ideas for the design of a program that could simulate this type of motion were generated during discussions with colleagues at the Educational Technology Center. However, the evolution of *Graphs and Tracks* from a preliminary stage to an award-winning program began at Carnegie Mellon[3], where the computing expertise and technology available at the Center for the Design of Educational Computing (CDEC) played a vital role. Development of the cT Programming Environment[4] greatly facilitated the task of simulating on the computer some of the laboratory activities that had proved successful in the *Kinematics* module.

The original program, as it was developed at Carnegie Mellon, has essentially the same structure as Part I (From Graphs to Motion) of the present version. Modifications and additions to Part I, as well as the design and implementation of Part II (From Motion to Graphs), were carried out at the University of Washington. In the discussion below, an example problem is used as the context for describing basic features of the program common both to the original version and to Part I. This description is followed by a brief summary of the improvements in Part I that resulted from continued development and intensive testing in the instructional setting provided by the Physics Education Group.

2.1.1 Description of Part I

Illustrated in Fig. 1 is one of the eight example problems that make up the core of Part I. For each problem, the computer presents a position versus time graph (shown in the figure by a dashed line), a diagram of a ball on a set of tracks, and a set of initial conditions. The student's task is to modify the track arrangement and initial conditions so that the ball will reproduce the motion represented by the graph.

---

[3] The current version of the program runs on a variety of Macintosh and IBM microcomputers.

[4] "cT" is a programming language. It is descended from TUTOR, the mainframe language used for PLATO at the University of Illinois. At Carnegie-Mellon, TUTOR was modified by Bruce Sherwood into CMU-TUTOR for use on workstations. The present version is called "cT."

As shown in Fig. 1, the five track segments are approximately the same length and are supported by six posts. The height of each post is represented by the number that appears on the screen below it. The slopes of the track segments can be altered by using the mouse to raise or lower the posts.

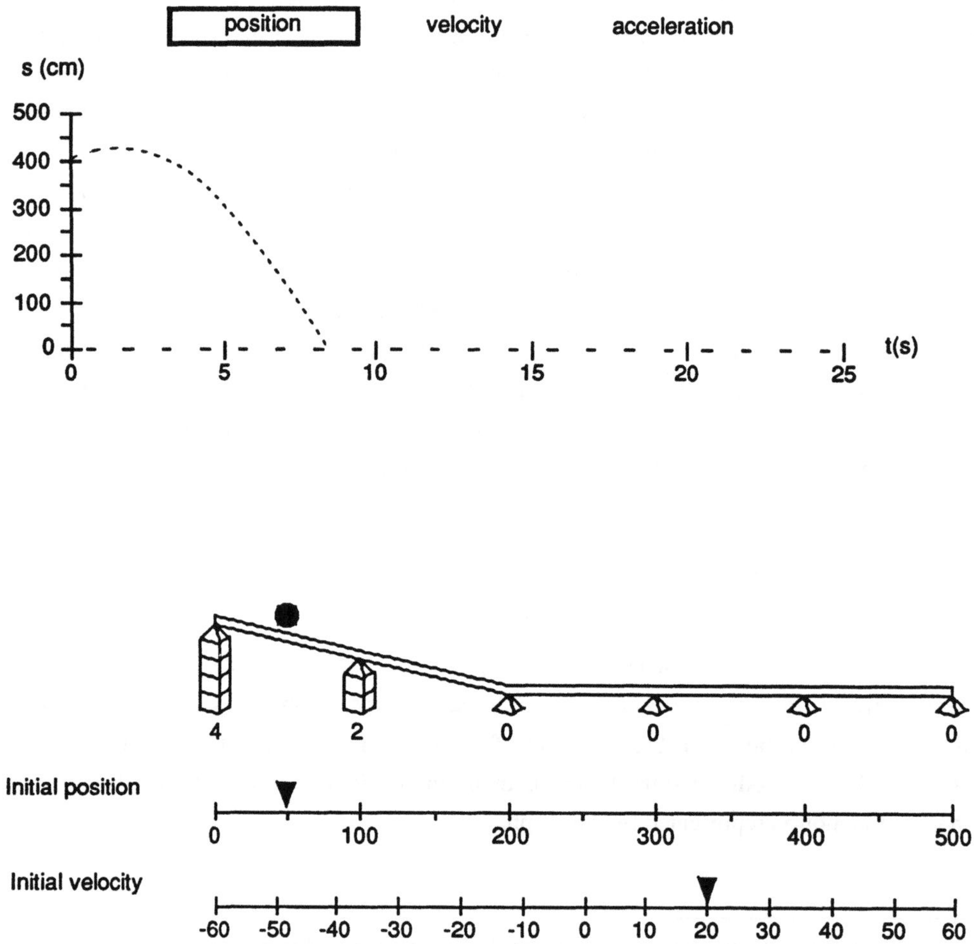

Fig. 1. Students are asked to rearrange the track segments and choose appropriate initial conditions so that the motion of the ball will reproduce the graph (Pt. I : FROM GRAPHS TO MOTION).

The program notes that the vertical scale is magnified and therefore the maximum height of any post is very small compared to the length of a segment. Like the track segments, the horizontal spacing of the posts appears to be uniform and the motion can be treated as essentially one-dimensional. Experience with more that 200 students has shown that the slight deviation from one-dimensional motion does not cause difficulty.

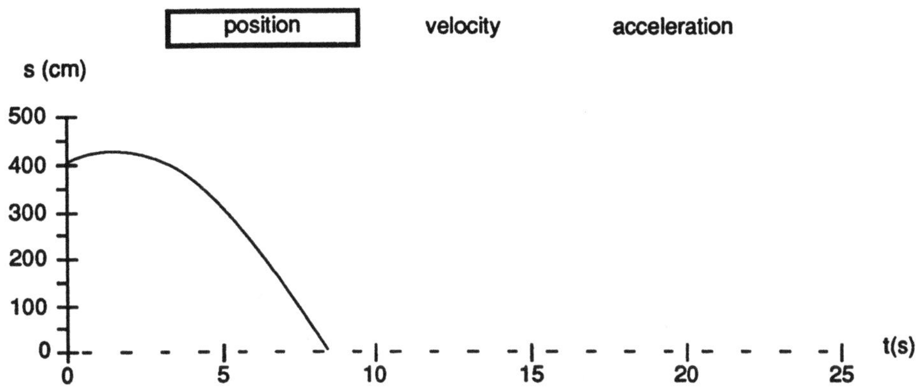

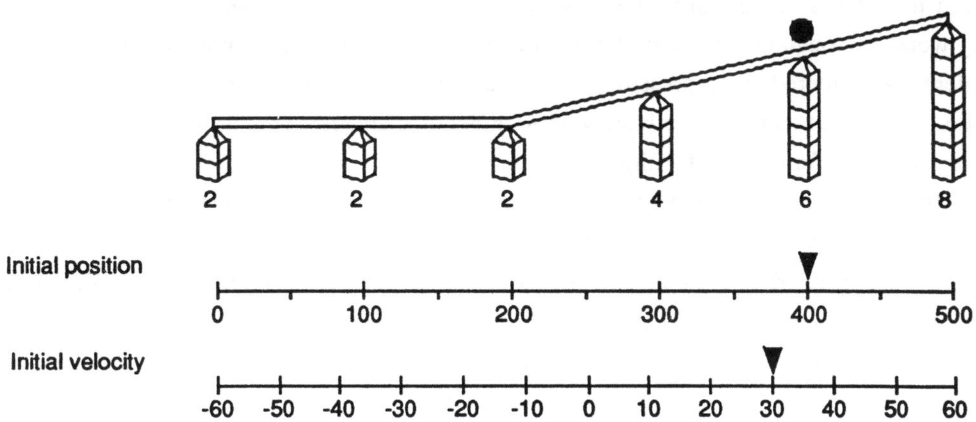

Fig. 2. The track arrangement and initial conditions that will result in the motion depicted by the graph shown in Fig. 1. (Pt. I: FROM GRAPHS TO MOTION).

In addition to adjusting the slope of the track, the student uses the mouse to select initial values for the position and velocity. These may be found by trial and error or read from the corresponding motion graphs. (The computer can be told to generate the velocity and acceleration versus time graphs that correspond to the given position versus time graph.) After adjusting the incline of the track and choosing the initial conditions, the student tells the computer to roll the ball. As the ball rolls, a solid curve is generated on the same set of axes as the given graph. The student compares the two graphs, modifies the track arrangement to obtain a better match, and repeats the motion as many times as desired. Shown in Fig. 2 is a track configuration that will reproduce the graph in Fig. 1. (Note that the dashed line in Fig. 1 is redrawn as a solid line when the correct motion is produced.)

2.1.2 Modifications based on testing in an instructional setting

As mentioned above, the initial development of *Graphs and Tracks* had been motivated by the investigation of student difficulties with motion graphs by the Physics Education Group. The instructional strategies in the *Kinematics* module that served as a model for the computer program had been based on findings from research and had been thoroughly tested in the classroom.[3,4] When these laboratory activities were implemented on the computer, intensive testing with students helped identify ways in which the program should be modified.

In the spring of 1987, the Physics Education Group began to use *Graphs and Tracks* as a means of probing more deeply into student understanding of motion graphs. Diane Grayson conducted a series of individual interviews in which she observed and questioned students as they worked through the program. The results provided insight into the types of problems the students encountered in progressing through the program. It became apparent that many students could not work productively without more guidance from the program and that thus there was a need for a "help" facility.

When Trowbridge returned to the University of Washington in the fall of 1987, we introduced *Graphs and Tracks* into the various introductory courses taught by the group and also into our more advanced courses for precollege teachers. Use of the program with large numbers of students helped identify programming errors, as well as areas needing more attention. Trowbridge modified the program by making changes and additions suggested by the interviews, by experience with students, and by members of the group.

The original program made no judgment about whether the graph generated was acceptable, nor did it provide any feedback to help the student decide which sections of the track were improperly arranged. The student was responsible for detecting any errors, determining what changes needed to be made, and deciding when the match was good enough to go on to another example. The major difference between Part I of *Graphs and Tracks* today and the original program is the presence of a help sequence that the student can request. If the graph generated by a proposed track arrangement fails to match the given graph, the help option can guide the student in searching for errors. The student's attention is directed, as appropriate, to the choices for initial position, initial velocity, or the slope of a particular section of the track. Systematic observation of students also suggested additional features that should be included in the program, such as labels for the relative heights of the posts. Other improvements include an introduction that provides instruction and practice on how to use the program and explicit confirmation by the computer when a student graph is correct. Instead of placing full responsibility on the student, as originally designed, the program assumes a greater tutorial role.

2.2 Development of Part II

On his return to the Physics Education Group, Trowbridge's attention was redirected to the research of the Physics Education Group on student understanding of motion graphs. Trowbridge decided to use the paper reporting on this work [2] as a guide in expanding *Graphs and Tracks* to include the process of going from tracks to graphs.

Part II (*From Motion to Graphs*) reverses the process of Part I (*From Graphs to Motion*). Instead of presenting the graph of a motion and asking the student to design an appropriate track arrangement, the program shows the ball rolling on a set of tracks and asks the student to sketch the corresponding motion graphs. As in Part I, an important factor that contributed substantively to the design of Part II was the earlier development of instructional materials by the Physics Education Group.[4] Activities that had proved successful in the laboratory provided models for the computer program.

The design of Part II posed challenges beyond those of Part I. The programming task was much more complex for both technical and intellectual reasons. Provision had to be made for allowing the student to draw graphs of varying shape and for enabling the computer to assess their correctness. The number of possible answers in Part II is much greater than in Part I, in which changes in shape are constrained by changes in incline of the track segments. Trowbridge solved this technical problem by modifying and incorporating into the program a drawing editor initially designed by Bruce Sherwood.

Important intellectual issues also had to be confronted in the design. It was necessary to program the computer to recognize a great variety of possible errors. The three graphs the student is expected to produce involve different variables, include lines with different slopes and curvatures, and must be consistent with one another. The computer must also be able to respond to a student who has made an error with an appropriate strategy that addresses the underlying conceptual difficulty. The role of research in helping address some of these problems is discussed after the brief description of Part II below. To a greater extent than Part I, Part II of *Graphs and Tracks* is an example of software development that has been strongly research-driven.

2.2.1 Description of Part II

There are eight example problems which make up the core of Part II. Each problem presents a particular track arrangement. One of the example problems, which is shown in Fig. 3, will be used in the description of Part II of the program.

The student begins the task of graphing the motion by asking the computer to roll the ball. As the ball rolls along the track, a verbal description is generated on the screen. The motion is as follows: the ball starts at a point part way up the incline, rolls up with decreasing speed until it reverses direction, increases in speed until it reaches the bottom of the incline, and then travels along the level segment of the track at a steady speed. The student is asked to draw position, velocity, and acceleration versus time graphs.

To draw the three graphs, the student may use the shapes given on the graphing palette shown in Fig. 3, or sketch freehand. The student can produce a continuous graph by choosing lines and parabolic curves from the graphing palette and then combining these elements with one another or with curves drawn by using the freehand option. The lines and parabolic curves represent motion on a section of track of constant slope. The curves can be clipped into smaller pieces. Thus it is possible to draw parabolic shapes not directly available from the graphing palette.

Feedback from the computer is available on request. When the student selects the question mark from the palette, the program assesses the correctness of the graphs and provides guidance to help the student rectify errors. Correct graphs for the given track configuration are included in Fig. 3.

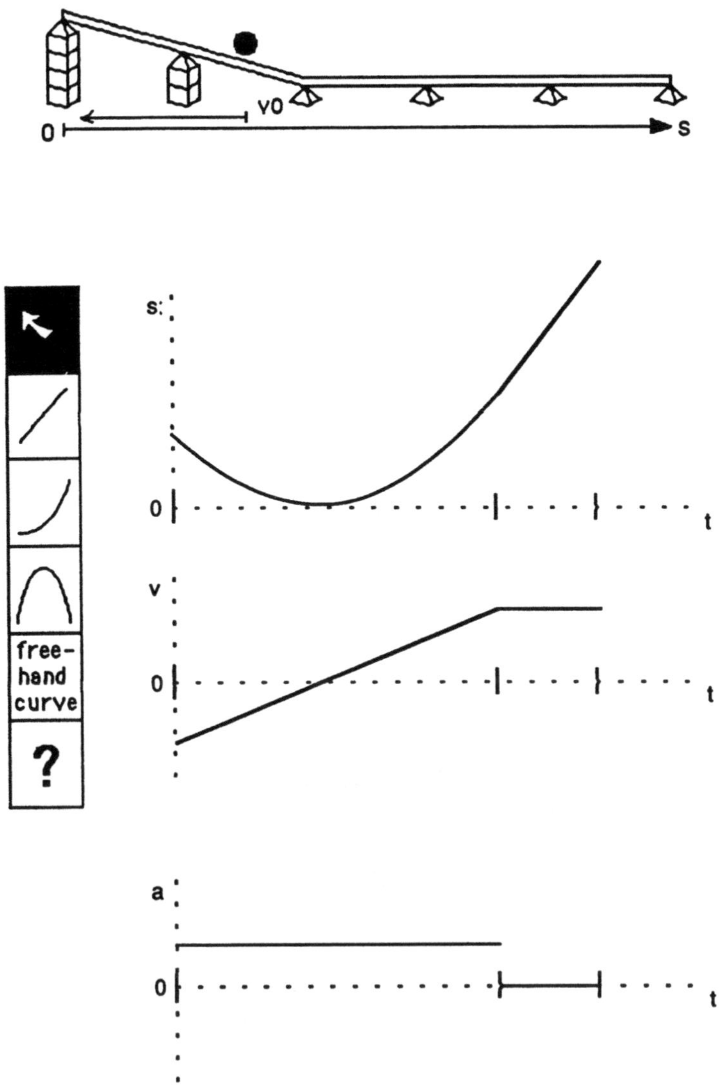

Fig. 3. A correct set of graphs for the given track configuration (Pt. II: FROM MOTION TO GRAPHS).

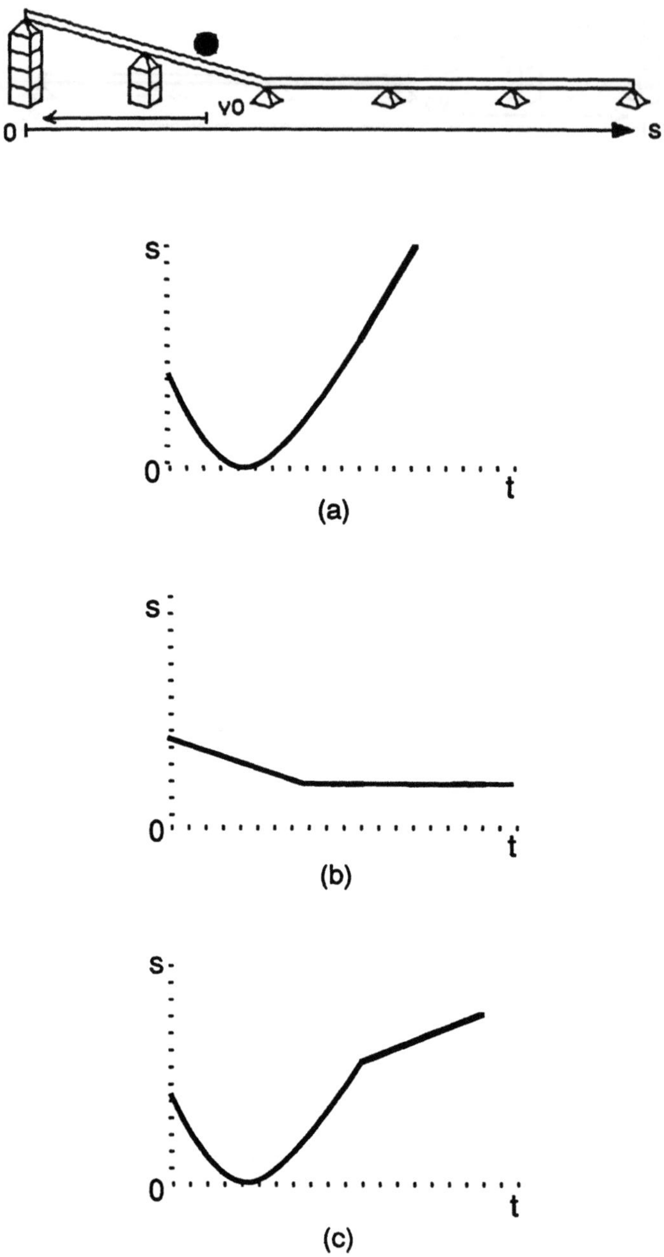

Fig. 4. A correct graph and two incorrect graphs that illustrate common errors for depicting the motion of the ball on the track arrangement shown. (a) Correct graph for the motion. (b) Incorrect graph resembling the shape of the track (c) Incorrect graph with kink connecting two segments (Pt. II: FROM MOTION TO GRAPHS).

2.2.2 Role of research in development of Part II

The descriptive study that provided the major motivation for the creation of Part II was conducted by the Physics Education Group over a period of several years.[2] The errors made by the several hundred students included in this research were classified into five groups. These may be summarized as difficulty in: (1) representing continuous motion by a continuous line, (2) separating the shape of a graph from the path of the motion, (3) representing a negative velocity on a v vs t graph, (4) representing constant acceleration, and (5) distinguishing among different types of motion graphs. Trowbridge used the results of this investigation to guide the design of Part II.

The role played by research in Part II is most easily conveyed by specific illustrations in the context of an example problem. Although it was not one involved in the study, the problem in Fig. 3 will be used for this purpose. Examples of specific difficulties identified in the research are described below in terms of the motion generated by the configuration shown. Each illustration is followed by a brief description of an instructional strategy used by the program to address the difficulty involved.

2.2.3 Examples of student difficulties and corresponding instructional strategies

When asked to plot a position versus time graph for a motion observed in the laboratory, students who had studied motion graphs in mathematics classes often could not sketch a graph that was qualitatively correct. For example, for the track arrangement shown in Fig. 4, the graph should resemble the one in Fig. 4(a). However, many students would sketch a graph, such as the one in Fig. 4(b), that resembled the shape of the track. In this case, the program would point out the failure of the graph to show the ball reversing direction and call attention to the fact that the position coordinate first decreases and then increases (or vice versa) when there is a turnaround.

An error at a more sophisticated level was made by students who drew a continuous straight line or curve for each segment of a motion but did not link the segments properly. For the example problem, this difficulty might lead to a graph such as the one in Fig. 4(c), which shows a kink instead of a smooth curve connecting the segments. A kink or cusp indicates an abrupt change in speed that does not occur in actual motion. The program would respond by calling attention to the kink and pointing out that the actual motion would change gradually and thus must be represented by a smooth curve.

When a negative velocity was involved, as is the case when an object reverses direction, many students could not represent the situation on a velocity versus time graph. The graph in Fig. 5 illustrates this kind of error. The reversal of direction should be indicated on this type of graph by a crossing of the time axis at the instant the ball reverses direction, not by a "V-shape." In addressing this error, the program would note that the graph as drawn does not show a change in the direction of the motion. All values of the velocity have the same sign.

As can be seen from the discussion above, students may produce graphs for a given motion that vary greatly in shape. The identification of specific difficulties through research allowed Trowbridge to design Part II to anticipate and confront common errors. Research also made it possible to design the program to permit students to make certain errors. In this way, latent difficulties that might otherwise pass undetected, could be evoked and therefore addressed.

3. Computer program drives research

The previous section described an approach to the design of computer-based instruction in which research in physics education plays a critical role. This section examines the interaction between research and computer-based instruction from a different perspective: the use of instructional software to identify and analyze conceptual difficulties. The computer is used as an investigatory tool to probe student understanding of the graphs of simulated moving objects.

In some preliminary studies, Diane Grayson conducted individual interviews in which computer tasks were used to examine student understanding in physics.[1] The procedures followed were similar to those we had used in individual demonstration interviews in other research. The main difference was that instead of observing or interacting with a physical system, the student viewed a computer screen and manipulated a keyboard, a mouse, or a motion probe.

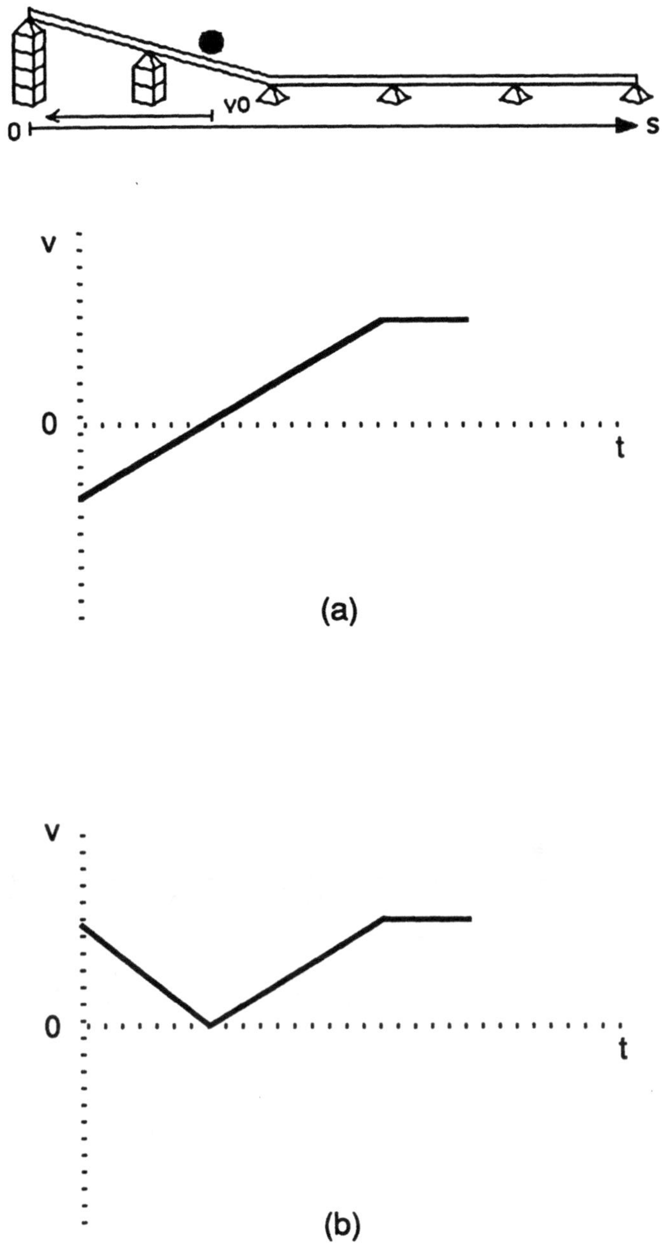

Fig. 5. A correct graph and an incorrect graph that illustrates common error for depicting the motion of the ball on the track arrangement shown. (a) Correct graph for the motion. (b) Incorrect graph in which a change in direction of the ball is represented by a V shape instead of a straight line through the horizontal axis (Pt. II : FROM MOTION TO GRAPHS).

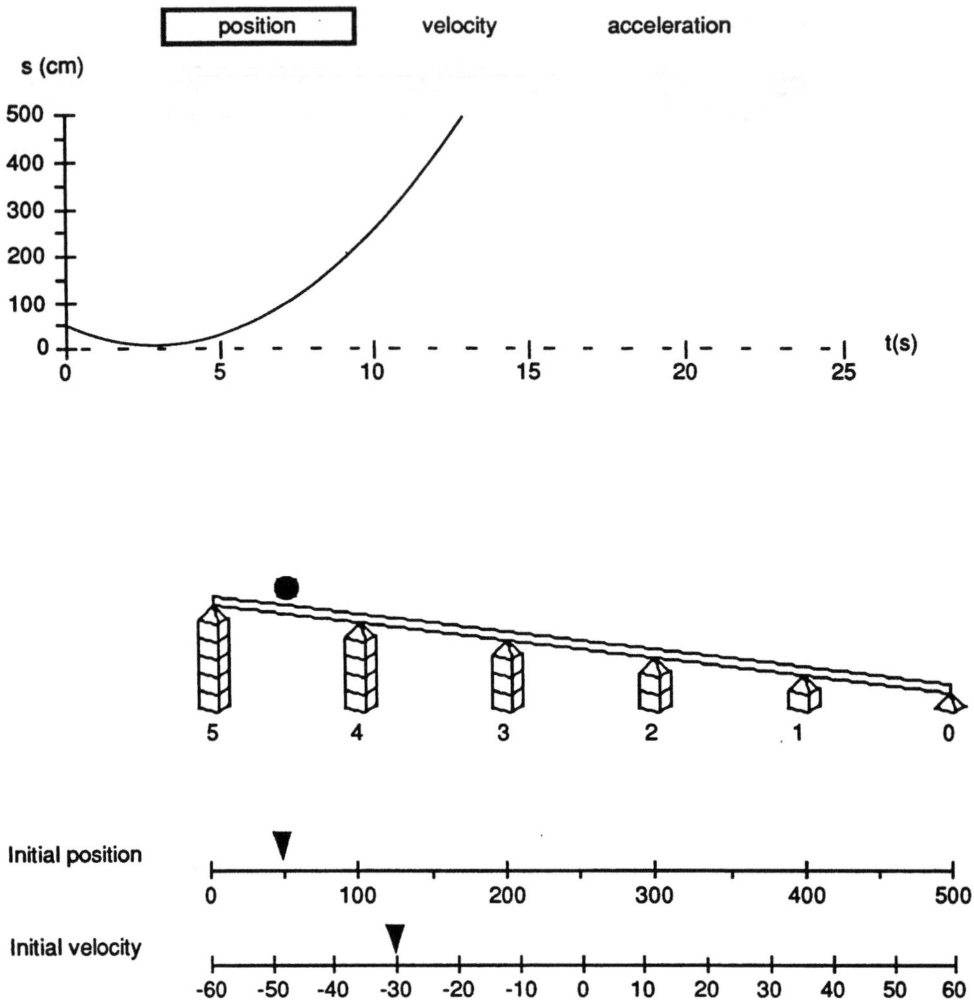

Fig 6. A track arrangement on which the ball will reproduce the motion depicted in the given graph (Pt; I: FROM GRAPHS TO MOTION).

An early version of *Graphs and Tracks* served as the basis of individual interviews in which the computer was used as an investigatory probe with students who had previously studied kinematics. The students were given several of the example problems from the program and asked to arrange the tracks so that the motion of the ball would generate the given graphs. In Fig. 6 is the graph contained in the first example given to the students, together with the correct track configuration and initial conditions needed to produce the graph. Initially, the ball must have a negative velocity, which diminishes gradually until the ball reverses direction. The velocity then becomes positive and continues to increase.

Nearly half the students (10 of 23) tried to make the ball move entirely in the positive direction. By inclining the first section of the track upward, as shown in Fig. 7, they tried to make the ball slow down but not turn around. With the range of initial conditions possible, however, the students could not give the ball a high enough velocity to continue moving forward. The resulting curve generated by the ball is shown in Fig. 7 at the lower left of the example graph. Some of the students who tried to make the ball move only in the forward direction were able to give a correct verbal description of the motion depicted in the example graph. However, they did not seem to recognize that although the graph extends toward the right along the horizontal axis (t), it reverses direction along the vertical axis (s). Hence, the motion must reverse direction in space.

For some students, the belief that the motion was always in the same direction was apparently so strong that they kept trying to increase the velocity as much as possible, even though the resultant graph bore no resemblance to the given graph.

The s vs t graph that was presented to the students in another task is a parabola that ends in a straight line, as shown by Curve A in Fig. 8. Curve A represents motion up and down a straight inclined track, ending on a level section. Many of the students were able to produce the track arrangement shown in Fig. 8, which can yield a curve of approximately the right shape. The initial conditions they chose, however, often resulted in a curve that was either too broad or too narrow. Curve B is an example of a graph that is too broad. About half of the students who wanted to make Curve B narrower tried to do so by increasing the initial velocity. They seemed to feel that the motion would be completed in a shorter time interval if the initial velocity were greater. However, increasing the velocity broadens the graph, as shown by Curve C in Fig. 8. The larger the initial velocity for a given incline, the farther the ball will travel up the incline and the longer it will take to turn around.

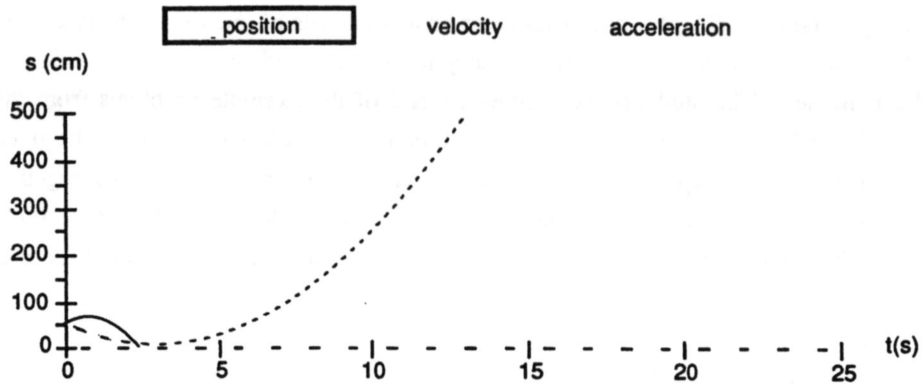

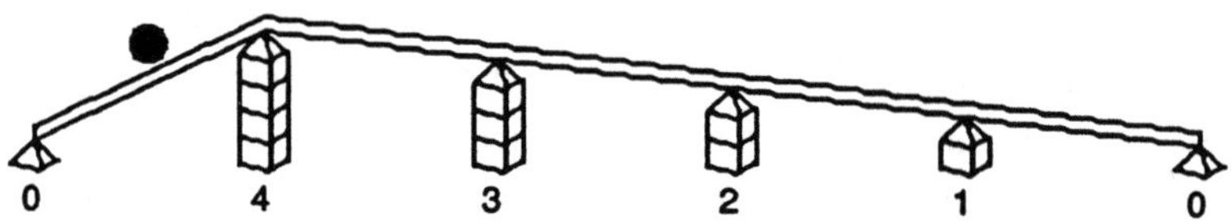

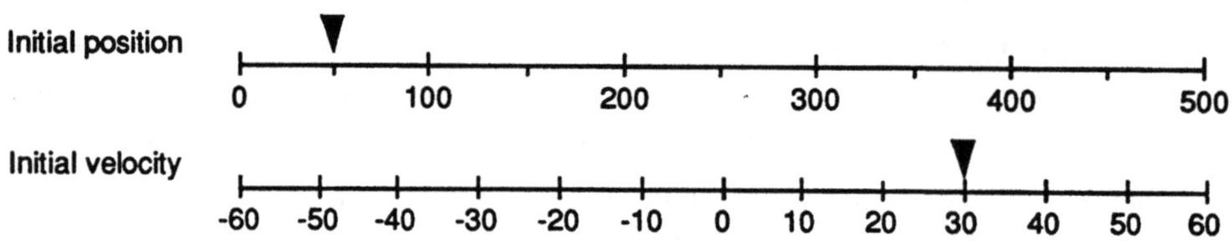

Fig. 7. An incorrect track arrangement produced by a student attempting to replicate the motion depicted in the graph in Fig. 6. The student tried to make the ball move only in the forward direction. (Pt; I: FROM GRAPHS TO MOTION).

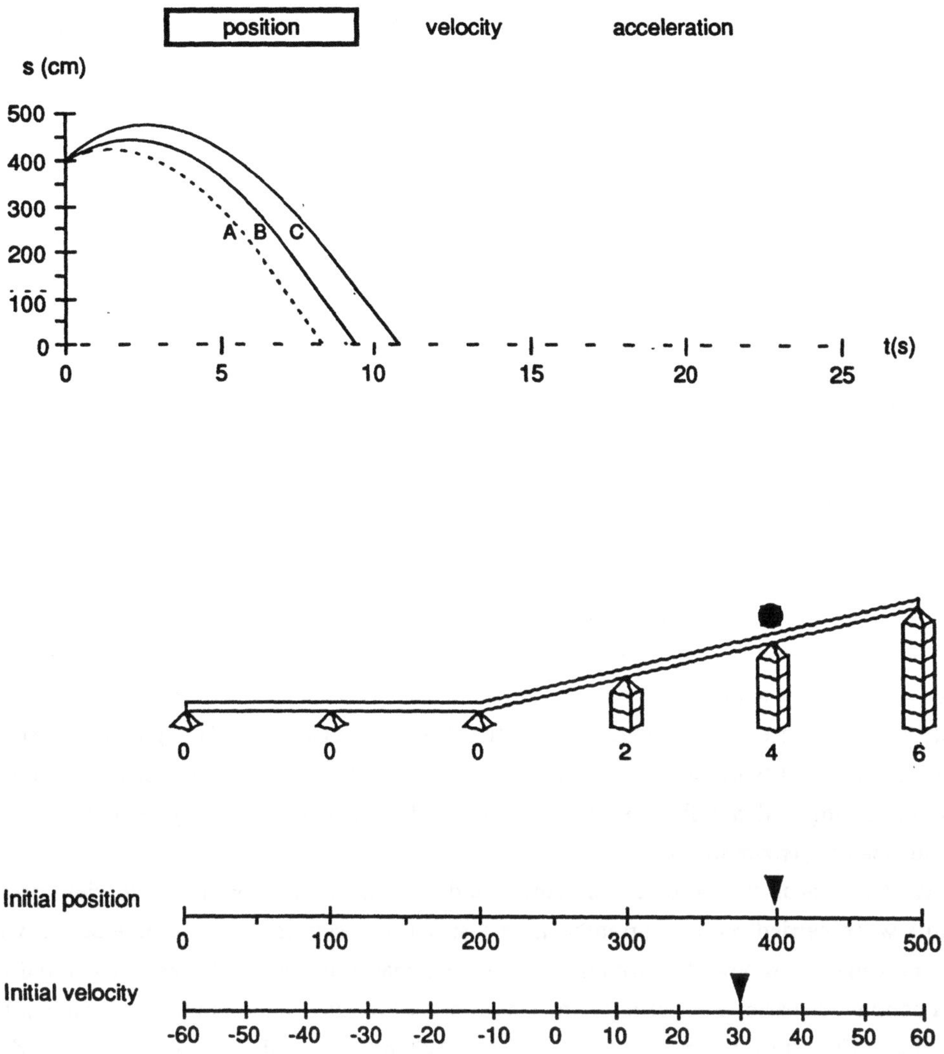

Fig. 8. An illustration of a common error made by students attempting to reproduce the motion depicted by the given graph (Curve A). Curve B results from the choice of an initial velocity that is too large. In trying to obtain a better match to the given graph, many students increased the initial velocity, thus broadening the graph still further as illustrated by curve C (Pt. I: FROM GRAPHS TO MOTION).

In the example illustrated in Fig. 8, use of the computer made it possible to examine aspects of conceptual understanding that are semi-quantitative in nature. Curves A, B, and C represent motions that are qualitatively similar, differing only in the magnitude of the initial velocity. Watching the procedures that students used in attempting to modify Curve B to match Curve A helped identify specific difficulties the students had in trying to relate changes in the shape of a graph to changes in the physical situation represented. This level of conceptual detail could not have been gleaned from written exercises or observed in the laboratory.

4. Conclusion

The evolution of *Graphs and Tracks* from its early beginnings to its present form is an example of how research can help guide the design of software. Experience with this program indicates that the computer can be an effective medium for helping students develop skill in the construction and interpretation of motion graphs. However, not all topics in physics are as suitable for computer-based instruction. There is a need for research both to identify appropriate areas of subject matter and to determine the kind of intelligent learning environment likely to be most effective for achieving a particular instructional goal. Throughout the process of development, it is necessary to take great care that the instructional materials that result address existing student difficulties and do not introduce new ones that may be particular to a specific learning environment.

We have found that research and software development are more likely to yield useful results when carried out concurrently in an instructional setting. Regular interaction with students can suggest new ideas for research; instructional strategies can be tested and modified on a continuous basis. Furthermore, the classroom serves as a daily reminder of some of the practical difficulties encountered in teaching. Whenever possible, implementation of an instructional strategy in an intelligent learning environment should take place only after the technique has been tried in the classroom or laboratory. Of course, there is no assurance that a strategy effective in one medium will be successful in another. Therefore, even when an approach has proved successful in the classroom, it is critical that testing with students continue throughout the entire process of development.

There is a danger that we may become so enthusiastic about what technology allows us to do that we lose sight of what students can or should do. What may appear to be a clever approach to the physics instructor or curriculum designer, may not work with students. As illustrated by the history of *Graphs and Tracks*, a great deal of effort and time may be required for the production of interactive software of high quality. Thorough planning and constant monitoring during development can help insure that we do not go astray in our efforts to develop instructional materials that are well-matched to the needs and abilities of students.

In addition to their use in instruction, intelligent learning environments may allow us to broaden our investigations into conceptual understanding beyond the possibilities provided by other media. Our experience suggests that computer-based interviews may be fruitful for investigating conceptual and reasoning difficulties on topics involving phenomena that cannot be directly observed in the laboratory. Results from such studies may have implications for the design not only of intelligent learning environments but also of traditional instructional materials.

Lastly and most importantly, there is a critical need to examine precisely what the student is learning from an intelligent learning environment. Is the student fully engaged intellectually, or only at the level required to develop skill at a video game? Does success on the given tasks indicate that the student has really understood the concepts and reasoning involved? Does learning extend beyond the medium itself so that the student can apply what he or she has learned on the computer to similar situations in other media or in the real world? Questions such as the foregoing must be considered in any meaningful evaluation of an intelligent learning environment.

Acknowledgements

The examples of research and software development used as illustrations in the discussion were based on the work of Diane Grayson and David Trowbridge. Other members of the Physics Education Group who have made substantive contributions in the preparation of this paper are Peter Shaffer and Mark Somers. The assistance of Betty Windham and Joan Valles is also gratefully acknowledged.

References

1. This paper draws on two earlier papers. McDermott, Lillian C.: "Research and computer-based instruction: opportunity for interaction," Am. J. Phys. **58**, 452-462 (1990); McDermott, Lillian C.: "Computers and research in physics education": In J. Risley and E. Redish (Eds.) *Proceedings of The Conference on Computers in Physics Instruction*, Reading, Mass.: Addison Wesley Publishing Co. (1990).
2. McDermott, Lillian C., Rosenquist, Mark L., and van Zee, Emily H.: "Student difficulties in connecting graphs and physics," Am. J. Phys. **55**, 503-513 (1987).
3. Rosenquist, Mark L. and McDermott, Lillian C.: "A conceptual approach to teaching kinematics," Am. J. Phys. **55**, 407-415 (1987).
4. Rosenquist, Mark L. and McDermott, Lillian C.: *Kinematics* (ASUW, University of Washington, Seattle, 1982).
5. Trowbridge, David E.: "Applying research results to the development of computer assisted instruction": In J. Risley and E. Redish (Eds.) *Proceedings of The Conference on Computers in Physics Instruction*, Reading, Mass: Addison Wesley Publishing Co. (1990).
   "Graphs and Tracks: An application of manipulable graphics," *Academic Computing*, pp. 24-25, 47-49 (May 1989).
6. Trowbridge, David E. and McDermott, Lillian C.: "Investigation of student understanding of the concept of velocity in one dimension," Am. J. Phys. **48**, 1020-1028 (1980); Trowbridge, David E. and McDermott, Lillian C.: "Investigation of student understanding of the concept of acceleration in one dimension," Am. J. Phys. **49**, 242-253 (1981).

Enhancing and Evaluating Students' Learning of Motion Concepts[1]

Ronald K. Thornton

Center for Science and Mathematics Teaching, Tufts University, Medford, Massachusetts 02155, USA

Abstract: Microcomputer-based laboratory (MBL) tools have been developed as an aid to all students, including the underprepared and underserved, in learning physical concepts. To guide this development, extensive work has been done to find useful measures of students' conceptual understanding that can be used in widely varying contexts. This article describes student learning of motion concepts by high school and college students in both traditional and MBL contexts. Students use MBL tools to collect physical data that are graphed in real time and then can be manipulated and analyzed. The MBL tools have made possible discovery-based laboratory curricula that embody results from educational research, allowing students to take an active role in their learning and encouraging them to construct physical knowledge from observation of the physical world. The curricula take advantage of the fact that MBL tools present data in an immediately understandable graphical form. They also encourage collaborative learning. The effectiveness of these methods compared to traditional high school and university methods for helping students learn basic motion concepts has been evaluated by pre- and post-testing and by observation. There is strong evidence for significantly improved learning and retention by students who used the MBL materials, compared to those taught in a traditional manner.

1. Introduction

Results from research in cognitive science and education substantiate the importance of basing development of scientific concepts and skills on concrete experience [7,15]. The "Tools for Scientific Thinking" project [17-20] at the Center for Science and Mathematics Teaching at Tufts University has developed microcomputer-based laboratory (MBL) tools and curricula that can help students make connections between the physical world and the principles which constitute scientific knowledge. The computer tools, which are being used in middle school, high school and college (including teacher preparation and enhancement programs), provide a convenient and effective means for students to collect and display physical data in a form that

---

[1] For footnotes and literature see References

NATO ASI Series, Vol. F 86
Intelligent Learning Environments
and Knowledge Acquisition in Physics
Edited by A. Tiberghien and H. Mandl

they can remember, manipulate, discuss and think about. The tools have enabled the development of curricula, based on research in science learning, that allow students to take an active role in their own learning. The primary purpose of these materials is to provide tools and techniques that enable students--including the underprepared, underserved, and science-anxious--to enhance their understanding of science concepts. Extensive work has been done to find useful measures of students' conceptual understanding which can be easily used in the widely varying contexts in which people learn science and which can be used with large numbers of students. After a brief introduction to microcomputer-based laboratory (MBL) tools and curricula, the paper will discuss student learning of motion concepts in traditional and non-traditional settings.

MBL tools of the style described in this article, were first developed at the Technical Education Research Centers (TERC) [2] for Apple II computers and are readily available [3]. For Macintosh computers, MBL tools able to measure force and motion simultaneously, current and voltage, light intensity, temperature, sound pressure, ionizing radiation and other physical properties have been developed by the Center for Science and Mathematics Teaching at Tufts University and by Dickinson College [4]. The probes are connected to the Universal Lab Interface which will connect to any computer with a serial port.

Students are not required to know anything about computers to use the MBL tools. Menu-driven, self-explanatory software is friendly, even for first time users, and encourages under-prepared and anxious students. With these tools, students are in control of their learning since they select the measurements to be made and the way the data are displayed. Data are displayed in digital and graphical form on the computer monitor as the measurements are taken. Students can transform and analyze the data, print graphs or tables, or save data to disks for later analysis. The tools do not simulate physical phenomena, but instead are a means of changing inexpensive computers into instruments for student-directed exploration of the physical world.

The following characteristics of these tools enhance the ability of students to learn in the contexts we have investigated. (1) The tools allow student-directed exploration but free students from most of the time-consuming drudgery associated with data collection and display. (2) The data are plotted in graphical form *in real time* , so that students get immediate feedback and see the data in an understandable form. (3) Because data are quickly taken and displayed, students can easily examine the consequences of a large number of changes in experimental conditions during a short period of time. The students spend a large portion of their laboratory time observing physical phenomena and interpreting, discussing, and analyzing data with their peers. (4) The hardware and software tools are --in general-- independent of the experiments. The variety of probes use the same interface box and the same software format. Students are able to focus on the investigation of many different physical phenomena without spending a large amount of time learning to use complicated tools (5). The tools dictate neither the phenomena to be investigated, the steps of the investigation, nor the level or sophistication of the curriculum. Thus a wide range of students from elementary school to university level are able to use this same set of tools to investigate the physical world.

2. Motion Detection

The motion detector and associated software is able to measure, display, and record the position, velocity, and acceleration of an object. The original motion detector was developed by TERC using a sonic transducer designed for Polaroid cameras. The motion probe (essentially a SONAR unit) transmits short pulses of high-frequency sound (50 kHz), then amplifies and detects the echo (much as a bat is able to do). The computer is programmed to measure the time between the transmitted and received pulses, and to calculate the position, velocity, and acceleration of the object causing the reflection. Any one of these quantities may be graphed on the screen as data are taken, and any one or more are available for display immediately after the measurements are completed. The motion detector can accurately detect objects between 0.5 and 6 m away. It detects the closest object in a roughly 15° cone.

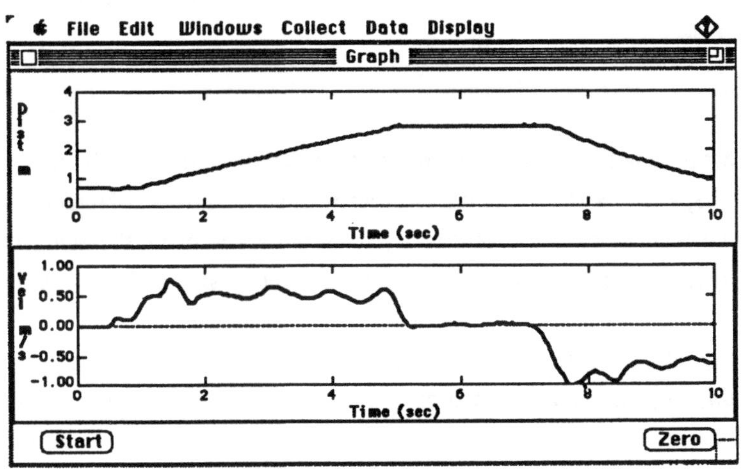

Fig. 1. Distance (position) and velocity graphs recorded on an Macintosh computer for a student walking away from the motion detector slowly, stopping, and then moving towards it more quickly. The velocity variations due to the student's steps show clearly in the velocity graph.

Figures 1 through 3 are examples of motion graphs produced by the detector interfaced to a Macintosh computer. The software enables students to change the scales of the vertical and time axes before or *after* the data are collected. Students who in the past plotted graphs in the corner of a large sheet of graph paper soon learn to make readable graphs--a general purpose skill, useful in many disciplines.

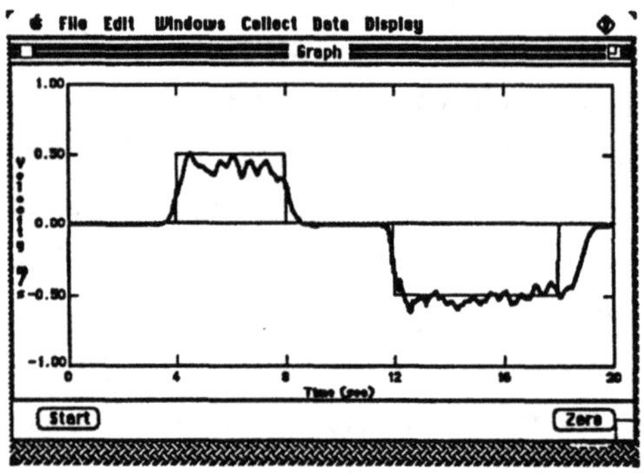

Fig. 2. Student's third attempt (heavy line) to duplicate with the motion of her body a velocity graph stored in an Macintosh computer (lighter line).

The software allows one set of data to be displayed on the screen while a new set of data are collected and graphed (perhaps after a slight change in experimental conditions). Numerical data are available in tabular form or can be read directly from the graph using the analysis software feature which presents digital values corresponding to the position of a movable cursor on the graph.

One of the most exciting features of the motion detector is its ability to detect and display graphs of the motion of *any* object. Thus, instead of using complex apparatus like nearly frictionless air tracks, which are not common to students' everyday experiences, the motion probe may be used to measure the motion of simple, common objects such as toy cars and even the motion of the students themselves. There is no other way of accurately displaying such graphs, certainly not in real time.

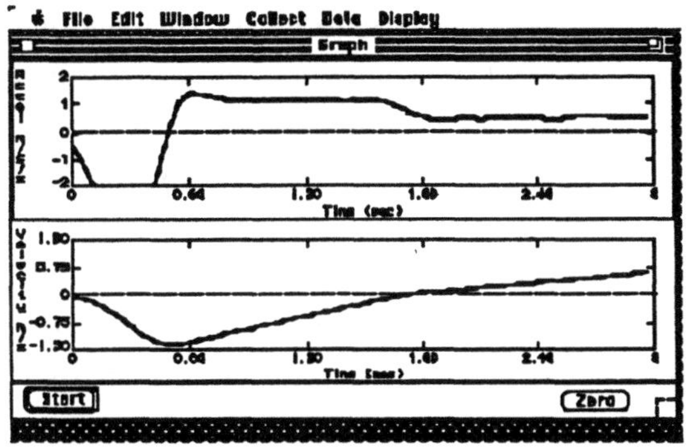

Fig. 3. Macintosh screen display of velocity and acceleration graphs for a cart (with friction) rolling up and back down an inclined ramp after a short push. (The motion detector was at the top of the ramp.)

3. The Motion Curriculum

These tools have made possible the Tools for Scientific Thinking curricula for university and secondary school students developed by the Center for Science and Mathematics Teaching at Tufts University [5]. These discovery-based laboratory curricula allow students to take an active role in their learning and encourage them to construct physical knowledge from actual observation. This article will discuss only the motion curriculum. The motion curriculum, in common with the others, makes substantial use of the results of educational research [12-14,21]. It uses a guided discovery approach and is intended for student groups of two to four. It supports the peer learning that is possible when data are immediately presented in an understandable form. It also uses predictions to engage the student and provide a vehicle for discussion. In addition, it pays attention to student alternative understandings that have been documented in the research literature and encourages students to construct knowledge for themselves. The introductory parts of the curriculum make substantial use of the students' own body motions to teach kinematic concepts.

The kinematics curriculum is divided into two pieces: *Introduction to Motion*, which covers concepts of position (distance from the motion detector) and constant velocities, and *Introduction to Motion--Changing Motion*, which covers changing velocities and acceleration. After some exercises with position graphs, the students work with the velocity of their own motion in *Introduction to Motion*. Students are asked to graph their velocities as they walk quickly and slowly toward and away from the motion detector. They are then asked questions about the graphical representations of *fast, slow, away from,* and *toward.* Figure 1 shows typical distance and velocity graphs for moving away from the detector slowly, stopping, and then moving toward it more quickly. Many students-- including university physics students-- are surprised, after viewing inclined distance graphs, to see roughly horizontal lines on the velocity graphs. These simple and quick exercises are designed to relate velocity graphs to various motions of real objects. Specifically, the exercises clarify the sign convention for velocities and firmly establish the relationship between an object's actual velocity and the displacement of the velocity graph from the time axis. Experience has shown that these simple exercises improve the conceptual understanding even of university students in calculus-based physics courses.

Students are asked to walk so as to duplicate a velocity graph that appears on the screen. (A similar exercise with a distance graph is included earlier in the curriculum.) Figure 2 shows the velocity graph to be matched (stored in the alternate display, so that it remains on the screen), and a student's third attempt to duplicate it with her own motion. The velocity graph is deliberately "unrealistic" (it shows several infinite accelerations) to provoke discussion. In addition, since the motion toward the detector is of longer duration than the motion away, students are forced to observe--when they run into the motion detector on the way back--that a velocity graph does not specify initial position.

Introduction to Motion--Changing Motion begins with a series of exercises designed to relate the sign of the acceleration to actual changing motions. Students are first asked to walk so as to produce graphs of distance, velocity, and acceleration for moving *away from* the detector while *speeding up*. After storing these in the alternate display, they produce graphs walking *toward* the detector quickly at first and then *slowing down*. Then, they are asked to predict what the graphs will look like for walking *toward* the detector, *speeding up*, and for walking *away, slowing down*. The students then move in the appropriate manner and compare their results to the predictions. Next they graph the motions of a cart, under various conditions. An interesting motion is that of a cart rolling up an inclined ramp, coming to rest and rolling back down. The velocity and acceleration graphs are shown in Figure 3. These graphs illustrate the advantages of tools that allow students to extend their observations beyond specialized cases such as uniform motion on a nearly frictionless air track. We have in fact used a toy car or a dynamics cart modified with an adjustable friction pad on the bottom, so that motion can be examined with different amounts of friction. The different slopes of the velocity graph on the way up and the way down, combined with the different heights of the acceleration graph, are convincing evidence for a frictional force that changes its direction when the cart reverses direction at the top.

The Tools for Scientific Thinking curriculum has been designed to be incorporated into traditional introductory physics courses found at most colleges and universities, where laboratory sections are often taught by teaching assistants with varying pedagogical skills, and where lecturers pay little attention to the laboratory. In place of the classroom discussions--

which under the best of circumstances would be used to consolidate the concepts learned in laboratory--each of these laboratories is accompanied by a homework assignment which the students complete after the laboratory session. They are asked to draw and interpret a number of graphs similar to and different from the ones they have produced in the laboratory. It is sometimes possible to have these problems discussed in discussion sessions. As shown below, high schools have much more flexibility to have pedagogically effective teacher-moderated student-focused discussions which improve student understanding.

4. Evaluating Student Learning of Motion Concepts in Traditional and MBL Environments

A visit to an MBL classroom/laboratory illustrates the contrast with a traditional class. Students are actively involved in their learning. They are sketching predictions and discussing them in groups of two or three. They are appealing to features of the graphs they have just plotted to argue their points of view with their peers. They are asking questions and, in many cases, either answering them themselves or finding the answers with the help of fellow students. There is a level of student involvement, success, and understanding that is rare in physics and physical science courses laboratory.

Student enthusiasm is wonderful, but are the MBL tools and curriculum really effective in helping students to learn motion concepts? Over the past 4 years we have been conducting studies of the effectiveness of traditional methods of teaching motion concepts and those curricula using MBL tools in middle school, high schools, colleges and universities in calculus and algebra-based introductory physics courses, and in teacher preparation and enhancement programs. Substantial work on curriculum development and evaluation was done at the college and university campuses that are part of the "Tools for Scientific Thinking" project [6]. In addition to co-authoring curricular materials, David Sokoloff of the University of Oregon has provided opportunities for extensive research on student learning [19]. Professors Sassi and Guidoni of the University of Napoli and Professor Borghi of Pavia have done research (some of which are reported in this publication) involving MBL with Italian university and high school students. The Workshop Physics Program [11], under the direction of Priscilla Laws at Dickinson College, has provided a more ideal college learning environment into which the tools and some of the curricular pieces have been adapted. In more usual environments, we have used pre- and post-testing and other forms of evaluation to examine the understandings of more than 1500 college and university physics students. We have also collected data for a large sample of secondary and middle school students. All of these contexts have provided strong evidence for significantly improved learning and retention by students who used the MBL materials, compared to those taught in lecture [17,19]. As examples of these learning results, I will discuss data from university calculus and algebra-based physics courses, high schools, and teacher preparation programs.

Velocity-Time Graphs

An object can move in either direction along the + distance axis. Choose the correct velocity-time graph(s) for each of the following questions. You may use a graph more than once or not at all. If you think that none is correct, answer J.

C 1. Which velocity graph shows the object moving away from the origin at a steady (constant) velocity? (A)

G 2. Which velocity graph shows the object standing still? (C)

D 3. Which velocity graph shows the object moving toward the origin at a steady (constant) velocity? (B)

F 4. Which velocity graph shows the object reversing direction? (E)

A 5. Which velocity graph shows the object increasing its speed at a steady (constant) rate? (not significant)

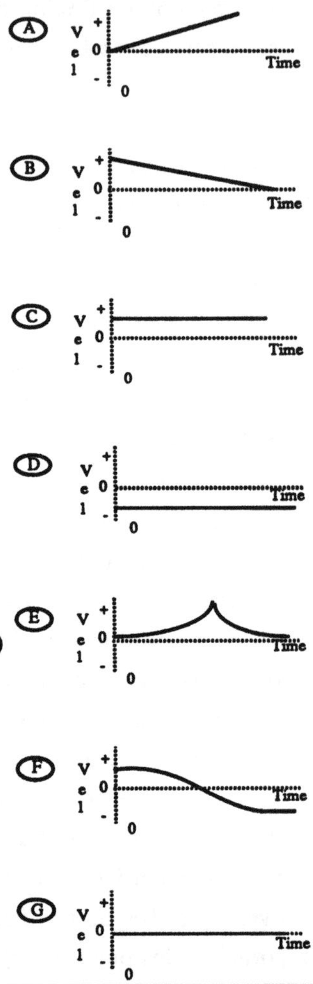

Fig. 4. Some of the multiple choice velocity questions asked on the kinematics pre- and post-tests. Questions 1 through 5 are the velocity questions referred to in the following figures. The most common wrong answer is shown in parentheses.

The pre- and post-tests that we have used in these studies consist in part of multiple choice questions. From earlier testing of students using free response questions requiring written answers and the drawing of graphs, we have constructed questions that seem to give a reasonable indication of students' basic knowledge of kinematics concepts and of graphical representation. Student results with these questions correlate well with their written answers on these and earlier tests. We find there are almost no random answers. Almost all students pick choices that we can associate with a small number of student models. This paper presents the results of only a few of more than 50 questions in different formats that are designed to distinguish among student models. Many of the multiple choice questions require students to choose the correct graph from a group of graphs. Testing on smaller samples shows that students who can pick the correct graph under these circumstances are almost equally successful at drawing the graph correctly without being presented with choices. The difficulties in convincing physics professors and high school teachers to give up course time for testing, our desire to make evaluation less subjective, and the effort involved in analyzing large samples moved us to use short answer questions for these studies.

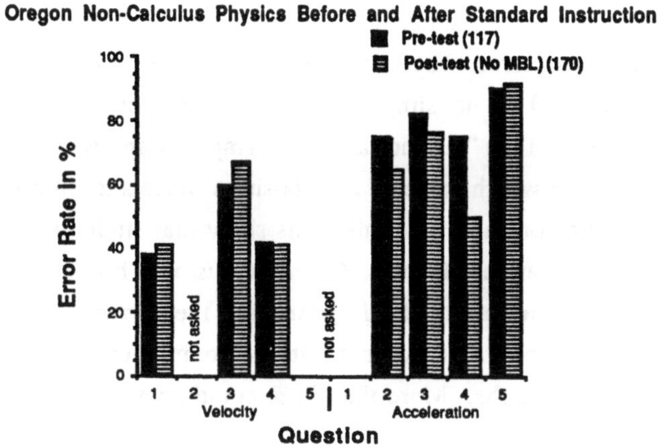

Fig. 5. Results for introductory physics lecture students (noncalculus) at the University of Oregon, Fall, 1988--comparison of student error rates on a few velocity and acceleration questions given before and after standard instruction.

Although a more complete understanding of student learning can be gained by an open-ended questioning process, we decided to use short answer questions in order to gather sufficient data at many different institutions to counter the common response that "my students do not have these difficulties you describe." From such questions we are also able to identify students with less common beliefs about motion and provide opportunities for open-ended responses.

Figure 4 shows a few of the simple, multiple choice velocity questions given on the pre-test, and Figure 5 shows the results of asking these questions (as well as the acceleration questions described below) of students in the non-calculus physics class at the University of Oregon. The post-test was given to all three sections, but it was only possible to give the pretest to two of the lecture sections. Since the populations of these lecture sections were random (the only selection criterion was time of day) the pretests should have been similar for all three. The results showed no significant differences between the two sections that were pretested. It was surprising to observe error rates as high as 40%-60% on these simple velocity questions *after kinematics had been covered in lecture* and students had done standard problems. Most physics professors had predicted that fewer than 10% of their students would miss these questions and felt that students who were unable to answer such simple questions understood very little kinematics. All of the lecturers were aware of the testing, and all made a special effort to teach kinematics graphing and concepts in their lectures. The large error rates on questions 1 and 3 (41% and 67%, respectively after instruction) are not simply the result of the wrong choice of sign. The most common error is the choice of the "distance analogs," graphs A and B. This is consistent with previous studies [13,21], in which students used position models to interpret velocity graphs. The different error rates on these two questions show that students have significantly more difficulty interpreting negative velocities. (This conclusion is borne out by the results of additional testing.) Neither the results of this pretest nor the correct answers were shared with the students. It should be noted that most students did not miss the questions because they were simply unable to read graphs. More than 90% could answer questions involving distance graphs correctly.

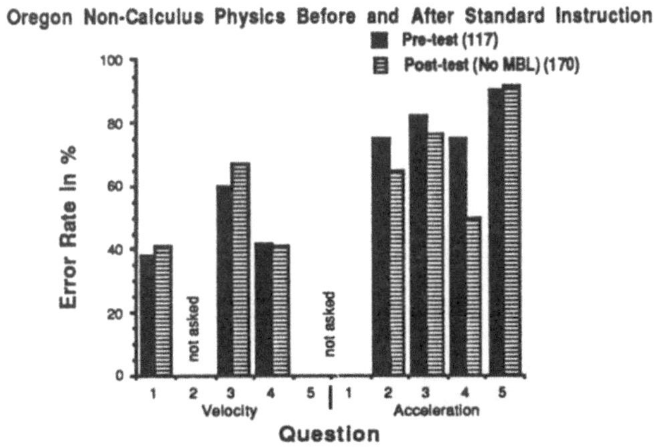

Fig. 5. Results for introductory physics lecture students (noncalculus) at the University of Oregon, Fall, 1988--comparison of student error rates on a few velocity and acceleration questions given before and after standard instruction.

In the fall of 1988, students in the introductory physics classes at Tufts University were asked the same questions as part of 50-question pre- and post-tests. The pretest was given *after* kinematics had been covered in lecture, but before the first MBL motion laboratory. The post-questions and the choices were shuffled on the post-test. The students turned in homework assignments that did not contain these questions. The results for students in the calculus lecture are shown in Figure 6. The results for calculus and non-calculus physics are remarkably similar [19]. The dark bars in Figure 6 are representative of the results of better students' understanding of these concepts after a good university calculus-based lecture course. The lighter bars are the result after the two MBL laboratories described above. (Velocity question number five only requires students to be able to "read" graphs and does not require understanding of velocity concepts. Most students answer this question correctly before instruction.)

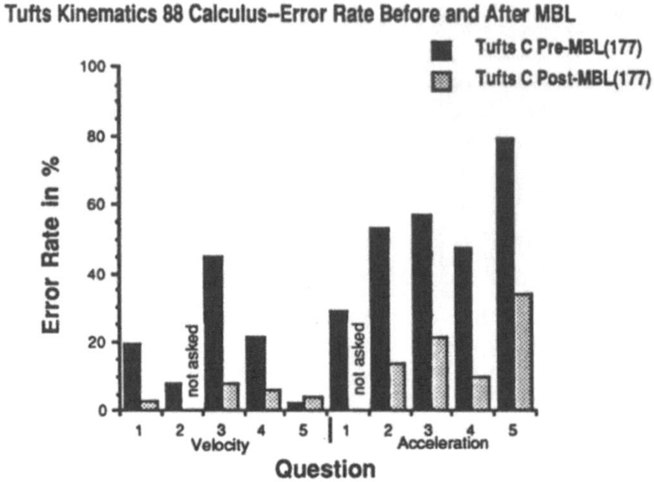

Fig. 6. Error rates on the same simple kinematics questions by Tufts University students taking introductory calculus-based physics. Dark bars are after lectures and problems while light bars are after two MBL laboratories.

Figure 7 shows some of the acceleration questions given to the same Tufts and Oregon students in 1988. Figure 6 shows the percentage of Tufts students in calculus physics courses who missed these questions. The pretest (given after lectures but before MBL) shows much higher error rates on the acceleration questions than on the velocity questions. These results are consistent with other studies [9,21] that show acceleration is considerably more difficult for students. Tufts students do not understand acceleration as well as they understand velocity after the MBL laboratories, but the improvement is still substantial. The Oregon results in Figure 5 also show that standard lecture instruction has little effect on student understanding of acceleration. Oregon students who take MBL laboratories show approximately the same improvement shown by the Tufts students [19]. As with velocity concepts, the results show that the students who only listened to lectures and did problems show almost no improvement on these concepts, while the students who worked with the MBL curriculum showed considerable improvement. A group of students at Oregon who had a special section instructor who consciously connected his lectures to the knowledge students discovered using the MBL curriculum [19] understood motion concepts even better. We have evidence that both university and high school students retain this knowledge with very little decrease in understanding over a period of months [17,19]. Some delayed post-tests have been given four months after the instruction with no substantial increase in error rates.

Acceleration-Time Graphs

A toy car can move in either direction along the + distance axis. Choose the letter of the acceleration-time graph which could correspond to the motion of the car described in each of the following.

You may use a graph more than once or not at all. If you think that none is correct, answer J.

_E_ 1. The car moves away from the origin at a constant velocity. **(B)**

_B_ 2. The car moves away from the origin, speeding up at a steady rate. **(A)**

_D_ 3. The car moves away from the origin, slowing down at a steady rate. **(C)**

_E_ 4. The car moves toward the origin at a constant velocity. **(D)**

_D_ 5. The car moves toward the origin, speeding up at a steady rate. **(C)**

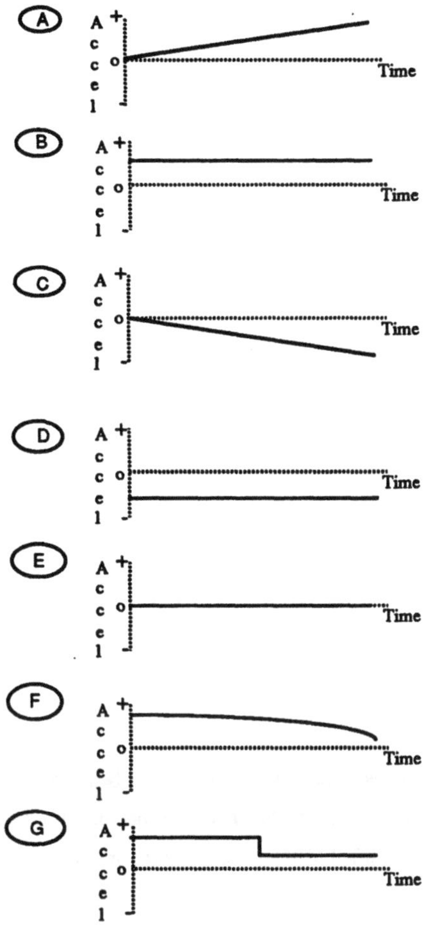

Fig. 7. Some of the multiple choice acceleration questions asked on the kinematics pre and post-tests. Questions 1 through 5 are the acceleration questions referred to in previous and following figures. The most common wrong answers are shown in parentheses.

Students preparing to be elementary school teachers in the US have had great difficulty understanding physics concepts in traditional courses. Professor Dewey Dyksta of Boise State University in Idaho offers a course which uses a modification of the MBL curriculum described above and uses methods of instruction based on constructivist methods to aid pre-service elementary teachers in learning physics. The students' understanding of kinematics concepts at the end of the course substantially exceeds those of calculus physics students after the usual instruction even though their understanding of such concepts at the beginning of the course is substantially less then students taking calculus physics (see Figure 8).

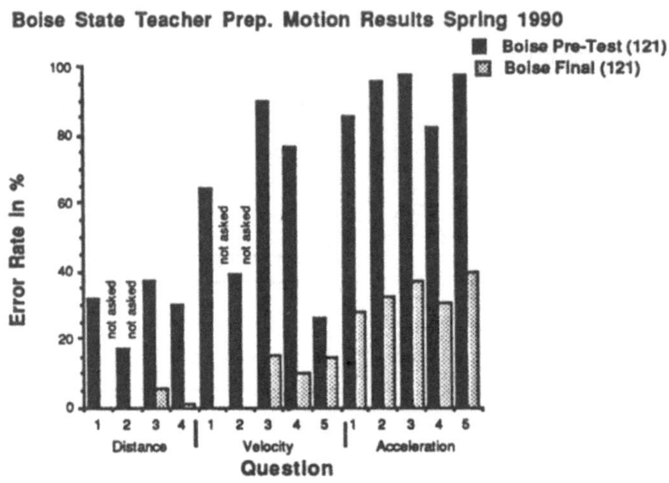

Fig. 8. Error rates on motion questions described above by pre-service elementary school teachers at Boise State University. Results at the beginning of the course are shown by the dark bars and results on the final exam after using a modified MBL motion curriculum are shown by the light bars.

High school students can also learn these concepts successfully using MBL. Figure 9 shows some typical error rates for a few hundred high school students from eight representative urban and rural/ suburban schools in Massachusetts. As before, the darker bars are error rates before any MBL curricula. In this case the students had no substantial instruction in motion from standard lectures or problems on kinematics before the pre-test data were taken. From this point of view the errors before instruction might be considered acceptable. However, results from the same schools in the previous year showed that traditional instruction resulted in error rates for the velocity questions only an average of 8% lower than those shown by the dark bars in Figure 9.

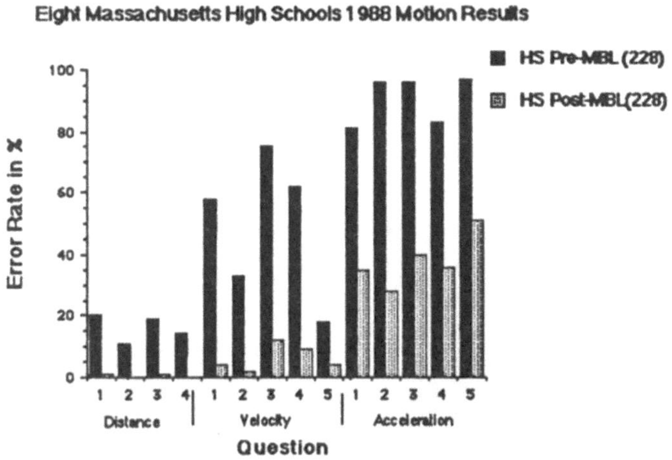

Fig. 9. Error rates on simple kinematics questions by 228 Massachusetts high school students before (dark bars) and after using the MBL curriculum (light bars).

The lighter bars show the error rates after using the MBL curriculum. The students also had no traditional kinematics instruction before the post test. The students do as well on the MBL curriculum without lectures as they do with lectures. The students were given about six to seven hours of MBL instruction and the material was the same as that covered in approximately two sessions of two to three hours at the university level. In spite of the large decrease in error rates for the acceleration questions, it is clear additional work needs to be done.

Part of the work may well be additional or more effectively moderated class discussion where students present the results of their working groups and discuss their ideas in the context of the entire class. Support for this position comes from the difference in student understanding at these Massachusetts schools between 1988 and 1987. In 1987 teachers were encouraged to let students talk about their ideas and results. This did not happen because the teachers felt the laboratories went so well and the students were so involved that no discussion of the results of the students' investigations, the questions presented in the curriculum, or the homework seemed necessary. They were happy with how and what their students learned. However, the student error rates were still approximately 30% on velocity questions 1 and 3.

In 1988, teachers did guide students in discussion of disagreements and difficulties, and the velocity error rates averaged 6%.

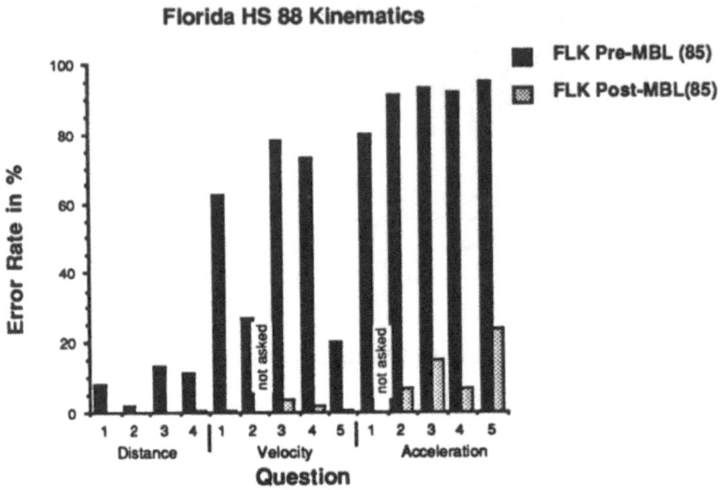

Fig. 10. Error rates on the same simple kinematics questions by 85 Florida high school students in three classes before (dark bars) and after using the MBL curriculum (light bars)

Three classes of students in a Florida high school, taught by a teacher in a national MBL group supported by the Center, used the same curriculum but participated in more substantial group discussions. These students have an average error rate of 11% on the acceleration questions after this experience. They started with an error rate of over 90% (Figure 10). The errors on the velocity questions are negligible.

These students understand kinematics better than most university students who take two two-hour MBL kinematics laboratories and seldom get the opportunity for general discussion or review except for that generated within their laboratory peer groups [17,19]. Students need an opportunity to consolidate their knowledge. A comparison of Figures 11 and 7 will show how well the high school students are doing compared to Tufts students taking calculus-based physics.

5. Conclusions

The MBL tools give students the opportunity to do *real science* in the introductory physics course. Thus students can experience the excitement of the process of science--the creative building and testing of models to explain the world around them. These tools give the science learner unprecedented power to explore, measure, and learn from the physical world. Because of their ease of use and pedagogical effectiveness, they make an understanding of physical phenomena more accessible to the naive science learner [16] and expand the investigations that more advanced students can undertake.

The tools, however, are not enough. Preliminary evidence shows that while the use of the MBL tools to do traditional physics experiments may increase the students' interest, such activities *do not* necessarily improve student understanding of fundamental physics concepts of the type discussed in this article. These gains in learning physics concepts appear to be produced by the combination of the tools *and* the appropriate curricular materials. In general, students improve their understanding of the physical concepts when they are guided by a curriculum to examine appropriate phenomena.

This article has presented evidence for enhanced learning of very basic physical concepts by groups of very different students using MBL tools and curriculum. We believe that the following five characteristics of the MBL learning environment--made possible by the tools, the curriculum, and the social and physical setting--are primarily responsible for the learning gains. Note that most of the characteristics of this learning environment bear more resemblance to the scientific workplace than to the usual educational environment.

1) Students focus on the physical world. Students learn concepts by investigating the physical world rather than only manipulating symbols or discussing abstractions as is common in traditional courses. In the MBL learning environment, actions in the physical world are directly linked to useful abstractions. For example, students who see the motion of their own bodies and of other objects displayed graphically in real time learn kinematics effectively.

2) Immediate feedback is available. The immediate feedback helps to make the abstract more concrete. The immediate coupling of the graphs to the physical phenomena seems to lead the students not only to understand graphing as a useful scientific symbol system, but also aids understanding of physical concepts when students are guided to examine appropriate phenomena. These observations are consistent with previous studies on a small number of students which suggested that even a short delay in the display of data in graphical form can reduce learning [8].

3) Collaboration is encouraged. Immediate feedback supports col[illegible]rative learning and collaborative work provides immediate feedback. Because da[illegible] presented in an understandable way, students can discuss the validity, the meaning, and the implications of the data with their peers. Learning is also enhanced by encouraging students to express their predictions and to discuss unexpected results with their peers. This process appears to be a powerful one in making students aware of alternative representations. The process of working collaboratively is closer to the way most scientists actually work.

4) Powerful tools reduce unnecessary drudgery. Instead of the time-consuming drudgery usually associated with data collection and display in the physics laboratory, student time is spent observing physical phenomena and analyzing and interpreting abstract representations of these phenomena (graphs). Students are able to concentrate more on discovering and understanding scientific concepts, and critical thinking skills are more easily developed. Hypothesis development and verification is encouraged by the ease and rapidity of repeating observations with changed experimental conditions. Powerful tools allow students to focus on authentic tasks in ways characteristic of scientists in the workplace. This is not commonly the case in school environments.

5) Students understand the specific and familiar before moving to the more general and abstract. The environment guides students to understand a specific, familiar (but often more complex) phenomenon before moving to the consideration of more general and abstract examples. Most students seem better able to understand motion when first considering, for example, their own motion (as complex as it is) as a reference point and then moving on to more idealized, less familiar (and less complex) motions with more general applicability such as frictionless motion or simple harmonic oscillation. Although it is difficult to abstract simple laws of physics from a complex, real process, grounding student understanding in the specific and familiar seems to make the abstract concepts more learnable. Moving from the specific to the general when investigating new concepts may also be more characteristic of the scientific workplace than the usual teaching and learning environment.

The evaluation of student conceptual learning in many different contexts has shown that curricula employing microcomputer-based laboratory tools allow students to develop a solid conceptual basis for understanding the world around them. Through the use of these tools, techniques and curricula, students connect their interactions with the physical world to the theories that constitute scientific knowledge.

References

1. This work was suppo[illegible] in part by the Fund for Improvement of Post-secondary Education (FIPSE) of the U.S. Department of E[illegible]ucation under the "Tools for Scientific Thinking" project and "Interactive Physics" at Tufts University, the National Science Foundation, the US Department of Education, the Board of Regents of Massachusetts, and Apple Computer, Inc.

2. Technical Education Research Centers, 1696 Massachusetts Avenue, Cambridge, MA 02138
3. The HRM Motion, Heat and Temperature and Sound microcomputer-based laboratory tools are available from Queue, Inc., 338 Commerce Drive, Fairfield, CT 06430.
4. For more information write to Ronald Thornton, Center for Science and Mathematics Teaching, Lincoln-Filene Building, Tufts University, Medford, MA 02155 and Prof. Priscilla Laws, Department of Physics and Astronomy, Dickinson College, Carlisle, PA 17013. These materials are available through Vernier Software, 2920 S.W. 89th Street, Portland, OR 97225.
5. The motion, heat and temperature and sound curricula were inspired by the middle school curriculum originally written at TERC.
6. Participating colleges and universities are Tufts University, University of Oregon, California Polytechnic State University, San Luis Obispo, Dickinson College, Massachusetts Institute of Technology, Muskingum College and Xavier University.
7. Arons, A.: Achieving wider scientific literacy, Daedalus 112, 91-117, 1983
8. Brassell, H.: The Effect of real-time laboratory graphing on learning graphic representations of distance and velocity, J. Res. Sci. Teaching 24, 385-395, 1987
9. diSessa, A.A.: The third revolution in computers and education, J. Res. Sci. Teaching 24, 343-367, 1987
10. Halloun, I.A. and Hestenes, D.: The initial knowledge state of college physics students, Am. J. Phys. 53, 1043-1055 (1985) and Common sense concepts about motion, Am. J. Phys. 53, 1056-1065, 1985
11. Laws, P.: Workshop physics: replacing lectures with real experience, Proc. Conf. Computers in Phys. Instruction, Redish, E. and Risley, J. eds. (Addison Wesley, Reading, MA, 1989), 22-32
12. McDermott, L.C.: Research on conceptual understanding in mechanics, Phys. Today 37, 24-32, July, 1984
13. McDermott, L.C., Rosenquist, M.L. and van Zee, E.H.: Student difficulties in connecting graphs and physics: Examples from kinematics, Am. J. Phys. 55, 503-513, 1987
14. Peters, P.C.: Even honors students have conceptual difficulties with physics, Am. J. Phys. 50, 501-508, 1982
15. Rosenquist, M.L. and McDermott, L.C.: A conceptual approach to teaching kinematics," Am. J. Phys. 55, 407-415, 1987
16. Thornton, R.K.: Access to college science: Microcomputer-based laboratories for the naive science learner, Collegiate Microcomputer V (1), 100-106, 1987
17. Thornton, R.K.: Tools for scientific thinking: Learning physical concepts with real-time laboratory measurement tools, Proc. Conf. Computers in Sci. Teaching, Redish, E. and Risley, J. eds. (Addison Wesley, Reading, MA, 1989), pp. 177-189
18. Thornton, R.K.: Tools for scientific thinking--microcomputer-based laboratories for teaching physics, Phys. Ed. 22, 230-238, 1987
19. Thornton, R.K. and Sokoloff, D.: Learning motion concepts using real-time microcomputer-based laboratory tools. Am. J. Phys. 58 (9), 858-66, Sept. 1990
20. Tinker, R.K. and Thornton, R.K.: Constructing student knowledge in science, in New Directions in Educational Technology, Scanlon, E. and O'Shea, T., eds. (Springer Verlag, Nato Science Series, to be published)
21. Trowbridge, D.E. and McDermott, L.C.: Investigation of student understanding of the concept of velocity in one dimension, Am. J. Phys. 48, 1020-1028 (1980) and Investigation of student understanding of the concept of acceleration in one dimension," Am. J. Phys. 49, 242-253, 1981

7 Technical Education Research Centers, 1696 Massachusetts Avenue, Cambridge, MA 02138.
8 The HRM Motion, Heat and Temperature and Sound microcomputer-based labs are available from Queue, Inc., 338 Commerce Drive, Fairfield, CT 06430.
9 For more information write to Ronald Thornton, Center for Science and Mathematics Teaching, Lincoln-Filene Building, Tufts University, Medford, MA 02155 and Prof. Priscilla Laws, Department of Physics and Astronomy, Dickinson College, Carlisle, PA 17013. These materials are available through Vernier Software, 2920 S.W. 89th Street, Portland, OR 97225.
10 The motion, heat and temperature and sound curricula [illegible] by the middle school curriculum [illegible].
11 Participating colleges and universities are Tufts University, University of Oregon, California Polytechnic [illegible]

List of Authors

NATO ASI Series F

Including Special Programmes on Sensory Systems for Robotic Control (ROB) and on Advanced Educational Technology (AET)

Vol. 1: Issues in Acoustic Signal - Image Processing and Recognition. Edited by C. H. Chen. VIII, 333 pages. 1983.

Vol. 2: Image Sequence Processing and Dynamic Scene Analysis. Edited by T. S. Huang. IX, 749 pages. 1983.

Vol. 3: Electronic Systems Effectiveness and Life Cycle Costing. Edited by J. K. Skwirzynski. XVII, 732 pages. 1983.

Vol. 4: Pictorial Data Analysis. Edited by R. M. Haralick. VIII, 468 pages. 1983.

Vol. 5: International Calibration Study of Traffic Conflict Techniques. Edited by E. Asmussen. VII, 229 pages. 1984.

Vol. 6: Information Technology and the Computer Network. Edited by K. G. Beauchamp. VIII, 271 pages. 1984.

Vol. 7: High-Speed Computation. Edited by J. S. Kowalik. IX, 441 pages. 1984.

Vol. 8: Program Transformation and Programming Environments. Report on a Workshop directed by F. L. Bauer and H. Remus. Edited by P. Pepper. XIV, 378 pages. 1984.

Vol. 9: Computer Aided Analysis and Optimization of Mechanical System Dynamics. Edited by E. J. Haug. XXII, 700 pages. 1984.

Vol. 10: Simulation and Model-Based Methodologies: An Integrative View. Edited by T. I. Ören. B. P. Zeigler, M. S. Elzas. XIII, 651 pages. 1984.

Vol. 11: Robotics and Artificial Intelligence. Edited by M. Brady. L. A. Gerhardt, H. F. Davidson. XVII, 693 pages. 1984.

Vol. 12: Combinatorial Algorithms on Words. Edited by A. Apostolico, Z. Galil. VIII, 361 pages. 1985.

Vol. 13: Logics and Models of Concurrent Systems. Edited by K. R. Apt. VIII, 498 pages. 1985.

Vol. 14: Control Flow and Data Flow: Concepts of Distributed Programming. Edited by M. Broy. VIII, 525 pages. 1985.

Vol. 15: Computational Mathematical Programming. Edited by K. Schittkowski. VIII, 451 pages. 1985.

Vol. 16: New Systems and Architectures for Automatic Speech Recognition and Synthesis. Edited by R. De Mori, C.Y. Suen. XIII, 630 pages. 1985.

Vol. 17: Fundamental Algorithms for Computer Graphics. Edited by R. A. Earnshaw. XVI, 1042 pages. 1985.

Vol. 18: Computer Architectures for Spatially Distributed Data. Edited by H. Freeman and G. G. Pieroni. VIII, 391 pages. 1985.

Vol. 19: Pictorial Information Systems in Medicine. Edited by K. H. Höhne. XII, 525 pages. 1986.

Vol. 20: Disordered Systems and Biological Organization. Edited by E. Bienenstock, F. Fogelman Soulié, G. Weisbuch. XXI, 405 pages.1986

Vol. 21: Intelligent Decision Support in Process Environments. Edited by E. Hollnagel, G. Mancini, D. D. Woods. XV, 524 pages. 1986.

NATO ASI Series F

Vol. 22: Software System Design Methods. The Challenge of Advanced Computing Technology. Edited by J. K. Skwirzynski. XIII, 747 pages. 1986.

Vol. 23: Designing Computer-Based Learning Materials. Edited by H. Weinstock and A. Bork. IX, 285 pages. 1986.

Vol. 24: Database Machines. Modern Trends and Applications. Edited by A. K. Sood and A. H. Qureshi. VIII, 570 pages. 1986.

Vol. 25: Pyramidal Systems for Computer Vision. Edited by V. Cantoni and S. Levialdi. VIII, 392 pages. 1986. *(ROB)*

Vol. 26: Modelling and Analysis in Arms Control. Edited by R. Avenhaus, R. K. Huber and J. D. Kettelle. VIII, 488 pages. 1986.

Vol. 27: Computer Aided Optimal Design: Structural and Mechanical Systems. Edited by C. A. Mota Soares. XIII, 1029 pages. 1987.

Vol. 28: Distributed Operating Systems. Theory und Practice. Edited by Y. Paker, J.-P. Banatre and M. Bozyiğit. X, 379 pages. 1987.

Vol. 29: Languages for Sensor-Based Control in Robotics. Edited by U. Rembold and K. Hörmann. IX, 625 pages. 1987. *(ROB)*

Vol. 30: Pattern Recognition Theory and Applications. Edited by P. A. Devijver and J. Kittler. XI, 543 pages. 1987.

Vol. 31: Decision Support Systems: Theory and Application. Edited bv C. W. Holsapple and A. B. Whinston. X, 500 pages. 1987.

Vol. 32: Information Systems: Failure Analysis. Edited by J. A. Wise and A. Debons. XV, 338 pages. 1987.

Vol. 33: Machine Intelligence and Knowledge Engineering for Robotic Applications. Edited by A. K. C. Wong and A. Pugh. XIV, 486 pages. 1987. *(ROB)*

Vol. 34: Modelling, Robustness and Sensitivity Reduction in Control Systems. Edited by R.F. Curtain. IX, 492 pages. 1987.

Vol. 35: Expert Judgment and Expert Systems. Edited by J. L. Mumpower, L. D. Phillips, O. Renn and V. R. R. Uppuluri. VIII, 361 pages. 1987.

Vol. 36: Logic of Programming and Calculi of Discrete Design. Edited by M. Broy. VII, 415 pages. 1987.

Vol. 37: Dynamics of Infinite Dimensional Systems. Edited by S.-N. Chow and J. K. Hale. IX. 514 pages. 1987.

Vol. 38: Flow Control of Congested Networks. Edited by A. R. Odoni, L. Bianco and G. Szegö. XII, 355 pages. 1987.

Vol. 39: Mathematics and Computer Science in Medical Imaging. Edited by M. A. Viergever and A. Todd-Pokropek. VIII, 546 pages. 1988.

Vol. 40: Theoretical Foundations of Computer Graphics and CAD. Edited by R. A. Earnshaw. XX, 1246 pages. 1988.

Vol. 41: Neural Computers. Edited by R. Eckmiller and Ch. v. d. Malsburg. XIII, 566 pages. 1988.

Vol. 42: Real-Time Object Measurement and Classification. Edited by A. K. Jain. VIII, 407 pages. 1988. *(ROB)*

NATO ASI Series F

Vol. 43: Sensors and Sensory Systems for Advanced Robots. Edited by P. Dario. XI, 597 pages. 1988. *(ROB)*

Vol. 44: Signal Processing and Pattern Recognition in Nondestructive Evaluation of Materials. Edited by C. H. Chen. VIII, 344 pages. 1988. *(ROB)*

Vol. 45: Syntactic and Structural Pattern Recognition. Edited by G. Ferraté, T. Pavlidis, A. Sanfeliu and H. Bunke. XVI, 467 pages. 1988. *(ROB)*

Vol. 46: Recent Advances in Speech Understanding and Dialog Systems. Edited by H. Niemann, M. Lang and G. Sagerer. X, 521 pages. 1988.

Vol. 47: Advanced Computing Concepts and Techniques in Control Engineering. Edited by M. J. Denham and A. J. Laub. XI, 518 pages. 1988.

Vol. 48: Mathematical Models for Decision Support. Edited by G. Mitra. IX, 762 pages. 1988.

Vol. 49: Computer Integrated Manufacturing. Edited by I. B. Turksen. VIII, 568 pages. 1988.

Vol. 50: CAD Based Programming for Sensory Robots. Edited by B. Ravani. IX, 565 pages. 1988. *(ROB)*

Vol. 51: Algorithms and Model Formulations in Mathematical Programming. Edited by S. W. Wallace. IX, 190 pages. 1989.

Vol. 52: Sensor Devices and Systems for Robotics. Edited by A. Casals. IX, 362 pages. 1989. *(ROB)*

Vol. 53: Advanced Information Technologies for Industrial Material Flow Systems. Edited by S. Y. Nof and C. L. Moodie. IX, 710 pages. 1989.

Vol. 54: A Reappraisal of the Efficiency of Financial Markets. Edited by R. M. C. Guimarães, B. G. Kingsman and S. J. Taylor. X, 804 pages. 1989.

Vol. 55: Constructive Methods in Computing Science. Edited by M. Broy. VII, 478 pages. 1989.

Vol. 56: Multiple Criteria Decision Making and Risk Analysis Using Microcomputers. Edited by B. Karpak and S. Zionts. VII, 399 pages. 1989.

Vol. 57: Kinematics and Dynamic Issues in Sensor Based Control. Edited by G. E. Taylor. XI, 456 pages. 1990. *(ROB)*

Vol. 58: Highly Redundant Sensing in Robotic Systems. Edited by J. T. Tou and J. G. Balchen. X, 322 pages. 1990. *(ROB)*

Vol. 59: Superconducting Electronics. Edited by H. Weinstock and M. Nisenoff. X, 441 pages. 1989.

Vol. 60: 3D Imaging in Medicine. Algorithms, Systems, Applications. Edited by K. H. Höhne, H. Fuchs and S. M. Pizer. IX, 460 pages. 1990.

Vol. 61: Knowledge, Data and Computer-Assisted Decisions. Edited by M. Schader and W. Gaul. VIII, 421 pages. 1990.

Vol. 62: Supercomputing. Edited by J. S. Kowalik. X, 425 pages. 1990.

Vol. 63: Traditional and Non-Traditional Robotic Sensors. Edited by T. C. Henderson. VIII, 468 pages. 1990. *(ROB)*

Vol. 64: Sensory Robotics for the Handling of Limp Materials. Edited by P. M. Taylor. IX, 343 pages. 1990. *(ROB)*

Vol. 65: Mapping and Spatial Modelling for Navigation. Edited by L. F. Pau. VIII, 357 pages. 1990. *(ROB)*

NATO ASI Series F

Vol. 66: Sensor-Based Robots: Algorithms and Architectures. Edited by C. S. G. Lee. X, 285 pages. 1991. *(ROB)*

Vol. 67: Designing Hypermedia for Learning. Edited by D. H. Jonassen and H. Mandl. XXV, 457 pages. 1990. *(AET)*

Vol. 68: Neurocomputing. Algorithms, Architectures and Applications. Edited by F. Fogelman Soulié and J. Hérault. XI, 455 pages. 1990.

Vol. 69: Real-Time Integration Methods for Mechanical System Simulation. Edited by E. J. Haug and R. C. Deyo. VIII, 352 pages. 1991.

Vol. 70: Numerical Linear Algebra, Digital Signal Processing and Parallel Algorithms. Edited by G. H. Golub and P. Van Dooren. XIII, 729 pages. 1991.

Vol. 71: Expert Systems and Robotics. Edited by T. Jordanides and B.Torby. XII, 744 pages. 1991.

Vol. 72: High-Capacity Local and Metropolitan Area Networks. Architecture and Performance Issues. Edited by G. Pujolle. X, 536 pages. 1991.

Vol. 73: Automation and Systems Issues in Air Traffic Control. Edited by J. A. Wise, V. D. Hopkin and M. L. Smith. XIX, 594 pages. 1991.

Vol. 74: Picture Archiving and Communication Systems (PACS) in Medicine. Edited by H. K. Huang, O. Ratib, A. R. Bakker and G. Witte. XI, 438 pages. 1991.

Vol. 75: Speech Recognition and Understanding. Recent Advances, Trends and Applications. Edited by P. Laface and Renato De Mori. XI, 559 pages. 1991.

Vol. 76: Multimedia Interface Design in Education. Edited by A. D. N. Edwards and S. Holland. XIV, 216 pages. 1992. *(AET)*

Vol. 77: Computer Algorithms for Solving Linear Algebraic Equations. The State of the Art. Edited by E. Spedicato. VIII, 352 pages. 1991.

Vol. 78: Integrating Advanced Technology into Technology Education. Edited by M. Hacker, A. Gordon and M. de Vries. VIII, 185 pages. 1991. *(AET)*

Vol. 79: Logic, Algebra, and Computation. Edited by F. L. Bauer. VII, 485 pages. 1991.

Vol. 80: Intelligent Tutoring Systems for Foreign Language Learning. Edited by M. L. Swartz and M. Yazdani. IX, 347 pages. 1992. *(AET)*

Vol. 81: Cognitive Tools for Learning . Edited by P. A. M. Kommers, D. H. Jonassen and J. T. Mayes. X, 278 pages. 1992. *(AET)*

Vol. 82: Combinatorial Optimization. New Frontiers in Theory and Practice. Edited by M. Akgül, H. W. Hamacher and S. Tüfekçi. XI, 334 pages. 1992.

Vol. 83: Active Perception and Robot Vision. Edited by A. K. Sood and H. Wechsler. IX, 756 pages. 1992.

Vol. 84: Computer-Based Learning Environments and Problem Solving. Edited by E. De Corte, M. C. Linn, H. Mandl and L. Verschaffel. XVI, 488 pages. 1992. *(AET)*

Vol. 85: Adaptive Learning Environments. Foundations and Frontiers. Edited by M. Jones and P. H. Winne. VIII, 408 pages. 1992. *(AET)*

Vol. 86: Intelligent Learning Environments and Knowledge Acquisition in Physics. Edited by A. Tiberghien and H. Mandl. VIII, 285 pages. 1992. *(AET)*

MIX
Papier aus verantwortungsvollen Quellen
Paper from responsible sources
FSC® C105338

If you have any concerns about our products,
you can contact us on
ProductSafety@springernature.com

In case Publisher is established outside the EU,
the EU authorized representative is:
Springer Nature Customer Service Center GmbH
Europaplatz 3, 69115 Heidelberg, Germany

Printed by Libri Plureos GmbH
in Hamburg, Germany